AQA
A Level Maths

A Level

Authors
Paul Hunt,
Steve Cavill, Rob Wagner

EXAM PRACTICE WORKBOOK

OXFORD
UNIVERSITY PRESS

Great Clarendon Street, Oxford, OX2 6DP, United Kingdom

Oxford University Press is a department of the University of Oxford.

It furthers the University's objective of excellence in research, scholarship, and education by publishing worldwide. Oxford is a registered trade mark of Oxford University Press in the UK and in certain other countries

British Library Cataloguing in Publication Data
Data available

978-0-19-841301-1

3 5 7 9 10 8 6 4

Paper used in the production of this book is a natural, recyclable product made from wood grown in sustainable forests.

The manufacturing process conforms to the environmental regulations of the country of origin.

Printed and bound by CPI Group (UK) Ltd, Croydon CR0 4YY

Although we have made every effort to trace and contact all copyright holders before publication this has not been possible in all cases. If notified, the publisher will rectify any errors or omissions at the earliest opportunity.

Links to third party websites are provided by Oxford in good faith and for information only. Oxford disclaims any responsibility for the materials contained in any third party website referenced in this work.

Contents

About this book

This book contains three sets of write-in, mock exam papers for the AQA A Level Maths exam (7367).

Full details of this exam specification can be for on the AQA website.

http://web.aqa.org.uk/7367

There are three papers in each exam set. Paper 1 covers Pure, paper 2 coves Pure and Mechanics and Paper 3 covers Pure and Statistics. All three papers are 120 minutes long and are each worth 100 marks.

The Large data set

The A Level examination will assume that you are familiar with a Large data set (LDS). In the exam, some questions will be based on the LDS and may include some extracts from it. It is AQA's intention that you should be taught using the LDS as this will give you a material advantage in the exam.

The LDS is part of the data set originally used to produce the UK government's report 'Family Food Statistic 2014' (DEFRA, 2015). The LDS contains data on household purchases of food and drink between 2001 and 2014, subdivided by government office region. An Excel spreadsheet containing the LDS is available from the AQA website at the address given above.

Answers

The back of this book contains short answers to all the questions.

Full mark schemes for each mock paper can be found online.

https://global.oup.com/education/content/secondary/series/aqa-alevel-maths/aqaalevelmaths-answers

Formulae

In the exam, you will be provided with a 'Formulae for A Level Mathematics' booklet that is for use in AS Level and A Level Maths qualifications. These are provided at the end of this book. The statistical tables for A Level Maths are also at the end of this book.

Calculators

All papers are calculator papers. Make sure that you know how to use your calculator, particularly for statistical functions. The rules on which calculators are allowed can be found in the Joint Council for General Qualifications document 'Instructions for conducting examinations' (ICE).

Name		Class	
Signature		Date	

Materials

You should have

- the booklet of formulae and statistical tables
- a graphical calculator.

Instructions

- Use a black pen for your working.
  Use a pencil for drawings.
- Answer **all** questions.
- Answer each question in the space provided for it; do **not** use the space provided for a different question. If you need extra space, ask for an additional answer book.
- All working should be inside the box drawn around each page.
- To avoid losing marks, show all necessary working.
- Include all rough working in this paper. If you do not want some work marked then cross it out.

Question	Mark
1	
2	
3	
4	
5	
6	
7	
8	
9	
10	
11	
Total	

Information

- Questions marks are shown in square brackets.
- There is a maximum of 100 marks available for this paper.

Advice

- Unless asked for a proof, you may quote any of the formulae in the booklet.
- You may not have to use all the answer space provided.

Answer **all** questions in the spaces provided.

1   Which of the following expressions have been simplified correctly?   **[1 mark]**

Circle your answer(s).

$$\frac{\cos^2 x - \sin^2 x}{\sin x} = \cos^2 x - \sin x \qquad \frac{3\cos x}{\cos x(\sin x \cos x - \cos^2 x)} = \frac{3}{\sin x - \cos x}$$

$$\frac{\cos^2 x - \sin^2 x}{\sin x} = 1 - 2\sin x \qquad \frac{3\cos x}{\cos x(\sin x \cos x - \cos^2 x)} = \frac{3}{\cos x(\sin x - \cos x)}$$

2   Evaluate $\dfrac{d}{dx}(e^{x^2})$ **[1 mark]**

Circle the correct answer.

$$e^{x^2+x} \qquad 2e^x \qquad e^{2x} \qquad 2xe^{x^2}$$

3   Find $\int \ln 2x \; dx$ **[1 mark]**

Circle the correct answer.

$$\frac{1}{x}+c \qquad\qquad x\ln x - x + c \qquad\qquad \frac{1}{2x}+c \qquad\qquad x\ln 2x - x + c$$

4   The diagram shows a sector ABC with radius r and angle θ, where θ is in radians.

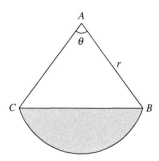

The arc length BC is P cm and the sector area ABC is Q cm²

It is given that $Q = 3P$

a   Find the length r **[3 marks]**

4 b   It is also given that the triangle *ABC* is equilateral.

Find the exact value of the area of the shaded segment.

You must fully justify your working.   **[5 marks]**

5   A quadrilateral *ABCD* is formed by joining the points of intersection of the lines with equations

$$y = 2x + 1 \qquad\qquad\qquad y - 2x = -10$$

$$2y = 1 - 4x \qquad\qquad\qquad y + 2x - 6 = 0$$

a   i   Write down the gradients of each of the four straight lines.   **[2 marks]**

_____

_____

_____

_____

_____

_____

ii   What can you deduce about the shape of the quadrilateral?

You must justify your answer.   **[2 marks]**

_____

_____

_____

_____

b   i   **Describe** how you would find the coordinates of the four vertices of the quadrilateral.

Do not include any calculations at this stage.   **[2 marks]**

_____

_____

_____

_____

_____

_____

_____

5 b ii Find the exact values of the coordinates of each of the four vertices of
the quadrilateral.   **[4 marks]**

_____

_____

_____

_____

_____

_____

_____

_____

_____

_____

c What are the lengths, in cm, of the shortest and longest sides of the quadrilateral?   **[4 marks]**

_____

_____

_____

_____

_____

_____

_____

_____

_____

_____

_____

6 a   Prove, by contradiction, that $\sqrt{2}$ is irrational.   **[6 marks]**

b   Simplify $\dfrac{\sqrt{2\sqrt{2\sqrt{2}}}}{\sqrt{2\sqrt{2}}}$, giving your answer in the form $\sqrt[a]{2}$, where a is an integer.   **[3 marks]**

6 c   A geometric sequence has first three terms 8, b and 4

    i   Find the exact value of b in the form $m\sqrt{n}$, where m and n are integers.   **[3 marks]**

    ii   Find the sum to infinity of the geometric sequence, showing all of your working.

      Give your answer in the form $f + g\sqrt{h}$, where f, g and h are integers.   **[5 marks]**

7 a Find the values of k such that $kx^2 + 4x + 5 = k$ has **no** real solutions.

You must show all your working. **[5 marks]**

b $(x-p)$ is a factor of $3x^2 - (p+8)x - (p+18)$, where p is an integer.

i Use the factor theorem to find the possible value(s) of p

You must show all your working. **[6 marks]**

7 b ii For each value of p found in part **b i**, find the other factor of the quadratic expression. **[2 marks]**

8 a Solve the equation $\dfrac{14}{x} - x = 5$

You must show each step of your working. **[3 marks]**

b Write down the equation of the line l shown in the diagram below. **[2 marks]**

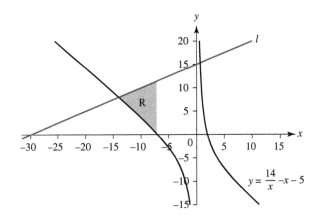

8 c By considering your answers to **a** and **b**, use calculus to find the area of the region labelled R.

Give your answer in the form $\dfrac{a}{b}+c\ln 2$, where a, b and c are integers.   **[11 marks]**

9 a Starting with the identity $\cos^2\theta + \sin^2\theta \equiv 1$, prove that $1 + \tan^2\theta \equiv \sec^2 x$ **[2 marks]**

b By using the identity from part **a**, or otherwise, solve the equation

$\sec^2\theta - \sec\theta = 1$ for $-\pi \leq \theta \leq 2\pi$

Give all values of θ in radians correct to three significant figures.

You must show every step of your working. **[5 marks]**

9 c Hence solve, for $-\pi \leq x \leq 2\pi$, the equation $\sec^2\left(\sin\frac{1}{2}x\right) - \sec\left(\sin\frac{1}{2}x\right) = 1$

Give your answers to three significant figures. **[4 marks]**

10 a Show that the equation $x^2 - 6 = 0$ has a root between $x = 2.4$ and $x = 2.5$ **[3 marks]**

10 b Hence, starting with $x_0 = 2.4$, use the Newton–Raphson method **once** to find an approximate value of $\sqrt{6}$ [3 marks]

c By defining a suitable function and then using the Newton–Raphson method (starting with $x_0 = 1.3$), find an approximation to the value of $\sqrt[7]{7}$ correct to six decimal places. [5 marks]

10 d

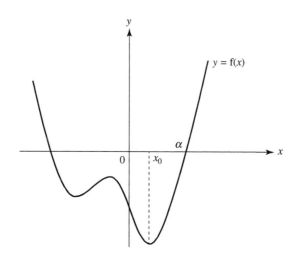

Jolene wishes to use the Newton–Raphson method to find an approximation to the positive root, α, of the curve $y = f(x)$ shown in the diagram.

She proposes starting the procedure at the point x_0, as shown in the diagram.

Will Jolene's method prove to be successful?

Justify your answer.   **[2 marks]**

11 a   Given that $x = \tan y$, find $\dfrac{\mathrm{d}x}{\mathrm{d}y}$ as a function of y **[1 mark]**

11 b Hence find $\dfrac{d}{dx}(\tan^{-1} x)$ as a function of x

You must fully justify your working.   **[4 marks]**

End of questions

Name _____ Class _____

Signature _____ Date _____

Materials

You should have

- the booklet of formulae and statistical tables
- a graphical calculator.

Instructions

- Use a black pen for your working.
  Use a pencil for drawings.
- Answer **all** questions.
- Answer each question in the space provided for it; do **not** use the space provided for a different question. If you need extra space, ask for an additional answer book.
- All working should be inside the box drawn around each page.
- To avoid losing marks, show all necessary working.
- Include all rough working in this paper. If you do not want some work marked then cross it out.

Information

- Questions marks are shown in square brackets.
- There is a maximum of 100 marks available for this paper.

Advice

- Unless asked for a proof, you may quote any of the formulae in the booklet.
- You may not have to use all the answer space provided.

Question	Mark
1	
2	
3	
4	
5	
6	
7	
8	
9	
10	
11	
12	
13	
14	
15	
16	
Total	

Section A

Answer **all** questions in the spaces provided.

1   If $y = x \ln 5$, what is $\dfrac{\mathrm{d}y}{\mathrm{d}x}$?

Circle your answer(s).

[1 mark]

$$\dfrac{x}{5} \qquad\qquad \ln 5 \qquad\qquad \dfrac{1}{5x} \qquad\qquad 5\ln x$$

2   The function $f(x) = x^3$ is mapped onto the function $g(x) = 8x^3$

Which of the following could describe this single transformation?

Circle your answer(s).   [1 mark]

Stretch parallel to the x-axis   Translation parallel to the x-axis

Stretch parallel to the y-axis   Translation parallel to the y-axis

3   By completing the square, find the radius and centre of the circle with equation

$$3x^2 + 3y^2 + 6x - 12y + 4 = 0$$

You must show all your working.   **[6 marks]**

4   Express $\dfrac{40-3x}{x^2(x^2-9x+20)}$ as the sum of partial fractions.

[8 marks]

5   The graph of $y = f(x)$ is shown for $-2\pi \leq x \leq 2\pi$

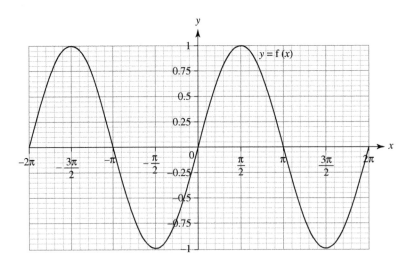

a   **i**   On the set of axes below, draw the graph of the gradient function, $y = f'(x)$

       Label the coordinates of the axes intercepts and the turning points.   **[2 marks]**

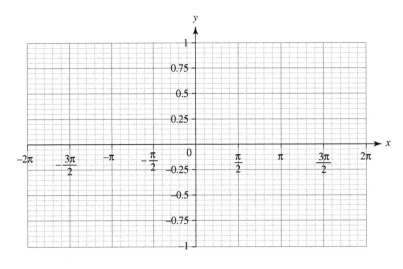

a   **ii**   Suggest a suitable equation for the graph of $y = f'(x)$ **[1 mark]**

5 b i On the set of axes below, draw the graph of the gradient of the gradient function, $y = f''(x)$

[2 marks]

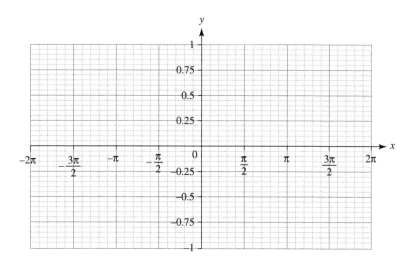

ii Suggest a suitable equation for the graph of $y = f''(x)$ **[1 mark]**

_____

_____

_____

_____

6 The square of the radius of a circle is plotted against its area.

This is then repeated for several other circles on the same set of axes.

The resulting graph shows a linear relationship, as seen below:

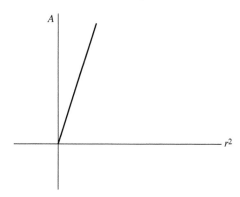

a What is the **exact** value of the gradient of the line? **[1 mark]**

_____

_____

_____

_____

_____

6　b　　Why is the graph only drawn in the first quadrant?　　　　　　　　　　　**[1 mark]**

c　　The circle shown below has equation $x^2 + y^2 = r^2$

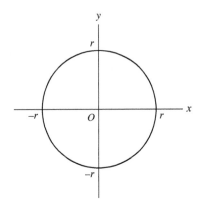

c　i　　By changing the subject to y and then using the substitution $x = r\sin\theta$, show that the area enclosed between the curve and the positive x- and y- axes is given by

$$r^2 \int_{0}^{\frac{\pi}{2}} \cos^2\theta \; d\theta$$

[6 marks]

6 c ii Evaluate the area enclosed between the curve and the positive x- and y-axes. **[4 marks]**

d Hence write down the area of the full circle. **[1 mark]**

7   The series below is the sum of a geometric progression and an arithmetic progression.

$32768 - 700 - 16384 - 740 + 8192 - 780 - 4096 - 820 + \ldots$

The series has 40 terms in total.

a   What is the 30th term of the series?   **[2 marks]**

b   What is the 37th term of the series?   **[2 marks]**

c   What is the exact value of the sum of all 40 terms of the series?   **[5 marks]**

8   The graph of $y = \log_a x$, where $a > 0$, is shown below.

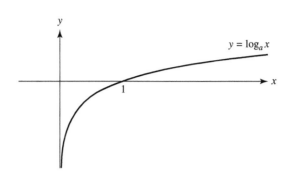

On the same set of axes, sketch the graphs of

a $y = \log_{2a} x$ **[2 marks]**

b $y = \log_{\frac{a}{2}} x$ **[2 marks]**

c $y = \log_{\frac{1}{a}} x$ **[2 marks]**

End of section A

Section B

Answer **all** questions in the spaces provided.

9   A box of mass 20 kg rests on a horizontal table.

What is the reaction of the table on the box?   **[1 mark]**

Circle your answer.

              20 N   2.04 N   196 N   9.8 N

10   A force of 25 N acts 1.3 m from a point A at an angle 72° to the horizontal line running through A, as shown in the diagram.

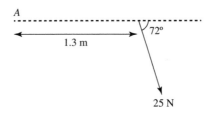

Calculate the moment of this force about A

You must show your working.   **[2 marks]**

11   A particle accelerates uniformly from u m s^{-1} to v m s^{-1} in t seconds.

The motion of the particle is shown in the velocity–time graph below.

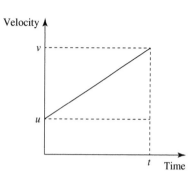

a i   Describe how you could use the graph to find the acceleration of the particle.   **[1 mark]**

_____

_____

_____

ii   Using your answer to **i**, show that $v = u + at$ **[3 marks]**

_____

_____

_____

_____

_____

_____

b   A car is initially moving at 50 km h^{-1}

It decelerates at a constant rate of 4 m s^{-2} until its velocity is 20 km h^{-1}

To the nearest metre, what distance does the car travel during this motion?   **[3 marks]**

_____

_____

_____

_____

_____

12   A uniform ladder of length 4.8 m and mass 20 kg stands on rough, horizontal ground and rests against a smooth, vertical wall. The ladder makes an angle of 70° with the ground.

a   In which direction does the wall exert a reaction force on the ladder?

Give a reason for your answer.   **[2 marks]**

b   Calculate the magnitude of the reaction force acting on the top of the ladder.   **[4 marks]**

13 A particle of mass 6 kg is acted on by forces of $(5\mathbf{i}+7\mathbf{j}+3\mathbf{k})\,\text{N}$, $(2\mathbf{i}-\mathbf{j}+6\mathbf{k})\,\text{N}$ and $(\mathbf{i}-\mathbf{k})\,\text{N}$

Find the magnitude of the acceleration of the particle. **[4 marks]**

14   A firework is projected from a point on horizontal ground.

The initial velocity of the firework is 150 m s^{-1} and it is projected at an angle θ to the horizontal.

There is a horizontal layer of thick cloud at a height of 0.5 km above the ground. In order for the audience to see the firework, θ must be chosen so that the path of the firework remains fully below the cloud.

a   Calculate the value of θ for which the missile *just* reaches the cloud.
You must show all your working.   **[3 marks]**

b   The technician makes a mistake and launches the firework when $\theta = 45°$

Assuming the firework does not explode, find, to the nearest tenth of a second, the time during which the firework is hidden by the cloud.

Show your working.   **[4 marks]**

15   A box of mass 5 kg is placed on a rough slope inclined at 40° to the horizontal.

It is released from rest and slides down the slope.

a   Draw a diagram to show the forces acting on the box.   **[2 marks]**

b   Find the magnitude of the normal reaction acting on the box.   **[2 marks]**

c   The coefficient of friction between the box and the slope is 0.3

Find the acceleration of the box.   **[4 marks]**

16   A particle moves in a horizontal plane, where **i** and **j** are perpendicular unit vectors.

At time t seconds the particle's velocity v m s^{-1} is given by $\mathbf{v} = 20\sin\left(\dfrac{\pi}{6}t\right)\mathbf{i} - 6\sqrt{t}\,\mathbf{j}$

a   Find an expression for the acceleration of the particle at time t seconds.   **[3 marks]**

b   The particle, which has mass 5 kg, moves under the action of a single force of magnitude **F** N

i   Find an expression for **F** in terms of t **[2 marks]**

ii   Show that, when $t = 9$, the magnitude of **F** is 5

You should also state the direction in which **F** acts.   **[4 marks]**

16 c   When $t = 9$ the particle is at the point with position vector $100\mathbf{i} + 100\mathbf{j}$

Find an expression for the position vector, $\mathbf{r}$ m, of the particle at time t **[6 marks]**

End of questions

| Name | | Class | |
| Signature | | Date | |

Materials

You should have

- the booklet of formulae and statistical tables
- a graphical calculator.

Instructions

- Use a black pen for your working.
  Use a pencil for drawings.
- Answer **all** questions.
- Answer each question in the space provided for it; do **not** use the space provided for a different question. If you need extra space, ask for an additional answer book.
- All working should be inside the box drawn around each page.
- To avoid losing marks, show all necessary working.
- Include all rough working in this paper. If you do not want some work marked then cross it out.

Information

- Questions marks are shown in square brackets.
- There is a maximum of 100 marks available for this paper.

Advice

- Unless asked for a proof, you may quote any of the formulae in the booklet.
- You may not have to use all the answer space provided.

Question	Mark
1	
2	
3	
4	
5	
6	
7	
8	
9	
10	
11	
12	
13	
Total	

Section A

Answer **all** questions in the spaces provided.

1   What is $\displaystyle\sum_{r=1}^{n} 1$?   **[1 mark]**

Circle your answer(s).

$$n \qquad\qquad 1 \qquad\qquad n+1 \qquad\qquad n!$$

2   What is the answer when the expression $x^8 - 1$ is fully factorised?   **[1 mark]**

Circle your answer(s).

$$(x^4+1)(x^4-1) \qquad (x^2-1)^4 \qquad (x^4+1)(x^2+1)(x+1)(x-1) \qquad (x^2+1)^2(x^2-1)^2$$

3 a On the same set of axes, sketch the curves $f(x) = |e^x - 4|$ and $g(x) = |e^x| - 4$

Label each curve with its respective equation.

Also, clearly indicate any points where either graph meets the coordinate axes. **[4 marks]**

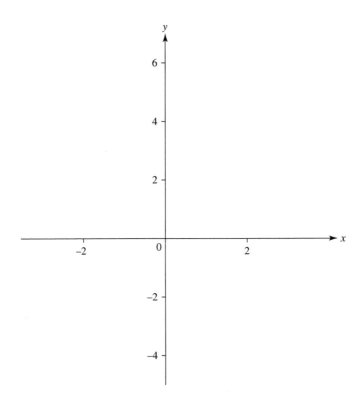

b Hence shade the region on your graph where $y \le |e^x - 4|$ and $y \ge |e^x| - 4$ **[1 mark]**

4 a Show that the distance, D, between the origin and a point (x, y) on the line $y = 4x + 3$ can be written as

$$D = \sqrt{17x^2 + 24x + 9}$$

[3 marks]

b i Use a method of calculus to find the exact values of x and y for which D^2 has a minimum value. You must justify that these values of x and y minimise D^2

[7 marks]

4 b ii Hence find the minimum distance between the line $y = 4x + 3$ and the origin. **[2 marks]**

_____

_____

_____

_____

_____

_____

5 a i $f(x) = \sin^{-1} x$ and $g(x) = \cos^{-1} x$

State the maximum domain, and corresponding range, of $f(x)$ and $g(x)$. **[4 marks]**

Domain $f(x)$ _____

Range $f(x)$ _____

Domain $g(x)$ _____

Range $g(x)$ _____

ii For the domains you identified in part **i**, sketch the graphs of $y = f(x)$ and $y = g(x)$ on the separate axes below.

On your sketches, mark the coordinates of any points where the graphs intersect the axes. **[2 marks]**

$y = \sin^{-1} x$

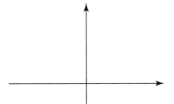

$y = \cos^{-1} x$

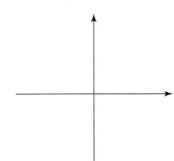

5 b   Using the substitution $u = \sin^{-1} x$, followed by a suitable identity, find the exact value of $\sin^{-1} x + \cos^{-1} x$ for any real x **[6 marks]**

c   Prove that the value of $\cos(2\sin^{-1} x)$ is $1 - 2x^2$ **[4 marks]**

6   The diagram below shows two concentric circles, C_1 and C_2

Circle C_1 has equation $x^2 + y^2 - 2x - 4y + 1 = 0$

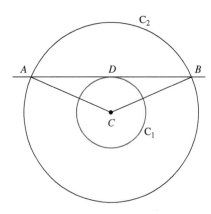

a   By completing the square, find the radius of circle C_1 and the coordinates of its centre, C **[3 marks]**

b   Circle C_1 has a horizontal tangent which meets the circle at D

i   Write down the equation of this tangent.   **[1 mark]**

6 b ii Given that circle C_2 has equation $x^2 + y^2 - 2x - 4y = 20$, find the exact length of the line segment AB

[5 marks]

iii Hence find the perimeter of the segment AB

Give your answer correct to 3 significant figures.

[6 marks]

End of section A

Section B

Answer **all** questions in the spaces provided.

7 To the nearest %, what is the probability that a value drawn from a Normal distribution is contained within one standard deviation of the mean? **[1 mark]**

Circle your answer.

38 68 95 99

8 *S* is a sample space for a random experiment. The probabilities for events *X* and *Y* follow this tree diagram.

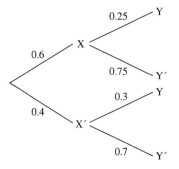

In the space provided below, draw a Venn diagram to show the same information. **[3 marks]**

9   A six-sided dice is rolled 270 times to see if it is a fair dice.
The following results are obtained.

Outcome	1	2	3	4	5	6
Frequency	41	35	46	72	31	45

Does this suggest that the dice is fair or not?
Justify your answer.   **[2 marks]**

10   A sports stadium checks all spectators' bags for security purposes as they enter the stadium.

The probability that a random person entering the stadium has brought a bag

is 0.3

The stadium has a capacity of 37 890 people.

a   Let the random variable X represent the number of bags which need to be checked by security at a single sporting event.

State the distribution of X, including any parameters.   **[1 mark]**

b   Explain why X can be reasonably approximated by a Normally distributed variable Y, and state the distribution of Y, including any parameters.   **[3 marks]**

10 c Use your random variable Y to show that the probability of the number of bags being checked between 11 000 and 11 500 (inclusive) is 0.9327 to 4 dp. **[4 marks]**

11 Events A, B and C are such that

$P(A)=0.5$, $P(B)=0.7$, and $P(C)=0.6$

$P(A\cap B\cap C)=0.13$, $P(A|C)=\dfrac{1}{3}$, $P(B|A)=0.78$, $P(B\cap C)=0.36$

The probabilities are to be represented on the Venn diagram below.

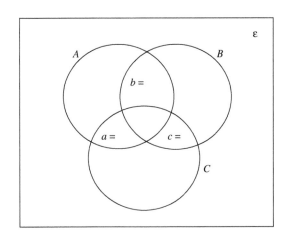

a Write the value of a on the Venn Diagram. **[1 mark]**

b Write the value of b on the Venn Diagram. **[1 mark]**

c Write the value of c on the Venn Diagram. **[1 mark]**

d Complete the Venn diagram by writing the correct probability in each blank space. **[3 marks]**

12   Raghav is studying the amount of fruit and vegetables that are eaten in England. He studies the average number of grams of fruit eaten, per person per week, over a 41-year period.

He obtains the following results.

	700–800	800–900	900–1000	1000–1100	1100–1200	1200–1300	1300–1400
Fruit	6	6	5	7	13	3	1
Vegetables	0	0	0	5	22	13	1

a   Is the data discrete or continuous?
Give a reason for your answer.   **[1 mark]**

b   Estimate both the median weight of fruit and the median weight of vegetables eaten per person per week.   **[3 marks]**

c   Estimate both the mean weight of fruit and the mean weight of vegetables eaten per person per week.   **[4 marks]**

12 d   Plot a frequency polygon for the amount of vegetables eaten per person per week.   **[3 marks]**

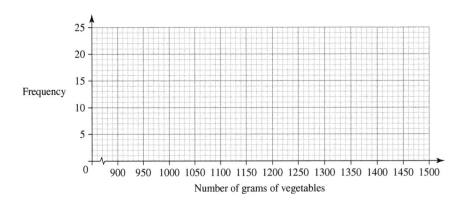

13   Amy is studying soft drink consumption in England.

She looks at the average volume of both low-calorie drinks and non low-calorie drinks, in ml, purchased per person per week over a period of 14 years.

She obtains the following results.

Year	2001	2002	2003	2004	2005	2006	2007	2008	2009	2010	2011	2012	2013	2014
Total	1698	1715	1869	1779	1657	1750	1642	1617	1627	1649	1604	1596	1604	1517
Not low calorie, n	1239	1249	1382	1359	1236	1245	1144	1152	1177	1104	930	875	836	757
Low calorie, l	459	465	487	420	421	505	498	465	450	545	674	720	769	760

a   For the **total** amounts of soft drink, calculate

i   The median,   **[1 mark]**

ii   The interquartile range (IQR).   **[1 mark]**

13 b i By referring to the Interquartile range (IQR), give a definition of an outlier. **[1 mark]**

_____

_____

_____

ii Determine if there are any outliers in the **total** amounts of soft drinks shown in Amy's results. **[2 marks]**

_____

_____

_____

Amy drew the following scatter diagram to show the consumption of **low-calorie** drinks against the consumption of **non low-calorie** drinks for each year that she studied.

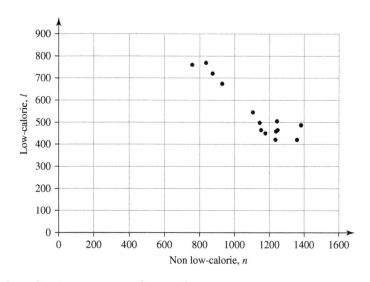

c Circle the plotted point corresponding to the year 2010 **[1 mark]**

d Amy calculates that the equation of the regression line of l on n is $l = 1226 - 0.607n$

i Draw Amy's regression line on the scatter diagram above. **[2 marks]**

_____

_____

ii Describe the correlation between the consumption of **low-calorie** drinks and the consumption of **non low-calorie** drinks shown by Amy's results. **[2 marks]**

_____

_____

13 d iii Use the regression line to predict the consumption of **low-calorie** drink per person per week when the consumption of **non low-calorie** drinks per person per week is 1 litre. **[2 marks]**

iv Give a reason why your prediction might not be accurate. **[1 mark]**

e Amy conducted her study to determine if there is any correlation between the amounts of low-calorie and non low-calorie soft drinks purchased per person per week.

i State Amy's hypotheses clearly. **[2 marks]**

ii The test statistic is −0.934 and the critical values are ±0.553 at the 5% level. Determine, in context, the conclusion to Amy's test. **[4 marks]**

End of questions

Name		Class	
Signature		Date	

Materials

You should have

- the booklet of formulae and statistical tables
- a graphical calculator.

Instructions

- Use a black pen for your working.
  Use a pencil for drawings.
- Answer **all** questions.
- Answer each question in the space provided for it; do **not** use the space provided for a different question. If you need extra space, ask for an additional answer book.
- All working should be inside the box drawn around each page.
- To avoid losing marks, show all necessary working.
- Include all rough working in this paper. If you do not want some work marked then cross it out.

Information

- Questions marks are shown in square brackets.
- There is a maximum of 100 marks available for this paper.

Advice

- Unless asked for a proof, you may quote any of the formulae in the booklet.
- You may not have to use all the answer space provided.

Question	Mark
1	
2	
3	
4	
5	
6	
7	
8	
9	
10	
11	
12	
Total	

Answer **all** questions in the spaces provided.

1   Which of the following pairs of inequalities describe the region labelled R?   **[1 mark]**

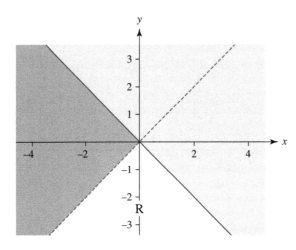

Circle your answer(s).

$x \leq y, x < -y$ $y \leq x, y < -x$ $x < y, x \leq -y$ $y < x, y \leq -x$

2   If the straight line $y = 2x - 3$ is reflected in the line $y = x$, what is the equation of the reflected line?   **[1 mark]**

Circle your answer.

$\frac{1}{2}x - 3$ $\frac{1}{2}x + 3$ $y = \frac{1}{2}(x - 3)$ $y = \frac{1}{2}(x + 3)$

3   If $y = (4x-3)^4(3-5x)^2$, show that $\dfrac{dy}{dx} = 6(3-5x)(13-20x)(4x-3)^3$

You should show all of your working clearly.   **[4 marks]**

If $y = (4x-3)^4(3-5x)^2$, show that $\dfrac{dy}{dx} = 6(3-5x)(13-20x)(4x-3)^3$

4   The graph of the function $y = \dfrac{2}{x}$ is sketched below.

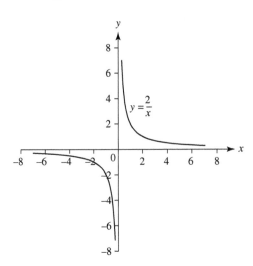

Sketch the following graphs on the axes provided below, clearly giving the coordinates of any points where your graphs intersect the coordinates axes, and the equations of any asymptotes.

a $y = \dfrac{2}{x+2}$ **[2 marks]**   **b** $y = -\dfrac{2}{x+2}$ **[2 marks]**

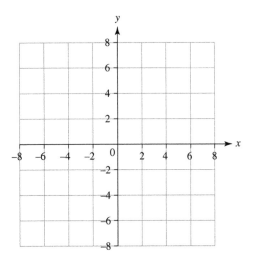   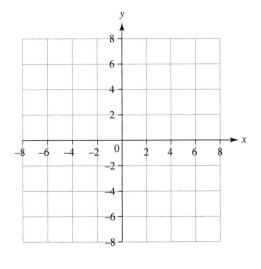

4 c $y = \dfrac{2}{|x+2|} - 2$ **[2 marks]**

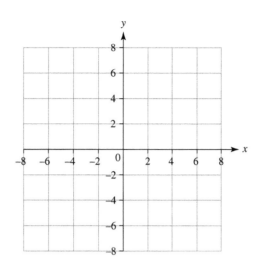

5 a   A curve, C_1, is defined by the parametric equations $x = \dfrac{t}{t^2+1}$, $y = \dfrac{1}{t^2+1}$

Show that the Cartesian equation of the C_1 is $x^2 + y^2 - y = 0$ **[4 marks]**

5 b Another curve, C_2, is defined by the parametric equations $x = \tan\theta + 3$, $y\cos\theta = 4$

Find a Cartesian equation for C_2 **[4 marks]**

6 Newton's law of cooling states that rate at which the temperature T of a body decreases over time t minutes is proportional to the difference between the temperature of the body and the temperature of its surroundings, T_0

a Explain how this leads to the first order differential equation

$$\frac{\mathrm{d}T}{\mathrm{d}t} = -k(T - T_0)$$

where k is a positive constant. **[2 marks]**

6 b A freshly brewed cup of oolong tea is set to rest at an initial temperature of 80 °C

The surrounding room temperature is a steady 20 °C

The tea takes 10 minutes to cool from 80 °C to 50 °C

i Find the value of the constant of proportionality in the form $a\ln b$, where a and b are rational numbers. **[8 marks]**

ii What will be the temperature of the oolong tea after it has been cooling for 20 minutes? **[3 marks]**

6 b iii How long will it take for the oolong tea to cool down to 30 °C?

Give your answer to the nearest second. **[3 marks]**

c What will happen to T as $t \to \infty$?

Why does your answer seem sensible? **[2 marks]**

7 a Daniel claims that

'64 is the only number between 1 and 1000 that is both a square and a cube number'.

Find a counter-example to show that Daniel's claim is incorrect. **[1 mark]**

7 b   Fiona states that:

'Every integer between 50 and 60 (inclusive) is the sum of at most 3 triangular numbers.'

Use the method of exhaustion to prove whether or not Fiona's statement is true.   **[3 marks]**

7   c   Prove by contradiction that there is an infinite number of prime numbers.   **[7 marks]**

8 a i Find $\int \ln x \, \mathrm{d}x$ [4 marks]

ii Hence show that $\int (\ln x)^2 \, \mathrm{d}x = x(\ln x)^2 + 2x(1 - \ln x) + d$, where d is an arbitrary constant. Show all of your working clearly. [4 marks]

8　b　Find the **exact** value of the area bounded by the curves $y = \ln x$ and $y = (\ln x)^2$, as shown shaded on the graph below. **[7 marks]**

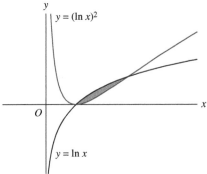

_____

_____

_____

_____

_____

_____

_____

_____

_____

_____

_____

_____

_____

_____

_____

_____

9 a Express the sum $\sin\theta+\sqrt{3}\cos\theta$ in the form $R\sin(\theta+\alpha)$,

where $R>0$ and $0<\theta<\dfrac{\pi}{2}$ **[3 marks]**

b Hence, find the **exact** solutions to the equation

$\sin 2\theta+\sqrt{3}(1-2\sin^2\theta)=2$ for $-2\pi<\theta<2\pi$

To gain full marks, you must show every step of your working. **[7 marks]**

10 a Find the binomial expansion of $(64 - x)^{\frac{1}{6}}$ up to and including the term in x^2 **[4 marks]**

b State the values of x for which the expansion in part **a** is valid.   **[2 marks]**

c Hence find an approximation to $\sqrt[6]{63}$, giving your answer to four decimal places.   **[2 marks]**

11 It is given that $y = \dfrac{x^{\frac{2}{3}}(6x+1)^4}{\sqrt{x^2+3}}$, where $x > 0$

a Show that $6\ln y = 4\ln x + 24\ln(6x+1) - 3\ln(x^2+3)$ **[4 marks]**

11 b Hence, by differentiating **implicitly**, show that

$$\frac{\mathrm{d}y}{\mathrm{d}x} = \frac{x^{\frac{2}{3}}(6x+1)^4}{\sqrt{x^2+3}}\left(\frac{2}{3x}+\frac{24}{6x+1}-\frac{x}{x^2+3}\right)$$

[5 marks]

12 a If $f(x) = \sqrt{x}$ and $g(x) = \sqrt[6]{x}$, find the composite function $fg(x)$

Give your answer in the form $fg(x) = x^{\frac{a}{b}}$, where a and b are integers. **[2 marks]**

b If $F(x) = x+1$, $G(x) = \dfrac{1}{x+1}$ and $H(x) = \dfrac{1}{x-1}$ find the composite function $FGH(x)$

Write your answer in the form $\dfrac{ax+b}{cx+d}$, where a, b, c and d are integers. **[4 marks]**

12 c $h(x) = 2x - 1$ and $k(x) = 2x^2 - 10x - 1$

Find a function $j(x)$ such that $hj(x) = k(x)$ **[3 marks]**

_____

_____

_____

_____

_____

_____

_____

_____

_____

_____

_____

_____

_____

End of questions

A Level Mathematics
Paper 2 (Set B)

AQA

Name	_____	Class	_____
Signature	_____	Date	_____

Materials

You should have

- the booklet of formulae and statistical tables
- a graphical calculator.

Instructions

- Use a black pen for your working.
  Use a pencil for drawings.
- Answer **all** questions.
- Answer each question in the space provided for it; do **not** use the space provided for a different question. If you need extra space, ask for an additional answer book.
- All working should be inside the box drawn around each page.
- To avoid losing marks, show all necessary working.
- Include all rough working in this paper. If you do not want some work marked then cross it out.

Information

- Questions marks are shown in square brackets.
- There is a maximum of 100 marks available for this paper.

Advice

- Unless asked for a proof, you may quote any of the formulae in the booklet.
- You may not have to use all the answer space provided.

Question	Mark
1	
2	
3	
4	
5	
6	
7	
8	
9	
10	
11	
12	
13	
14	
15	
Total	

Section A

Answer **all** questions in the spaces provided.

1 What is the minimum value of $5+\cos 3\theta$? **[1 mark]**

Circle your answer.

$$2 \qquad 4 \qquad 5 \qquad 8$$

2 Given that $3^{2x-1}=20$, what is the value of x? **[1 mark]**

Circle your answer(s).

$$\frac{\ln 60}{2\ln 3} \qquad \frac{1}{2}\ln 20 \qquad \frac{1}{2}\left(\ln\frac{20}{3}+1\right) \qquad \frac{1}{2}\ln\frac{20}{3}$$

3 If $\int_1^2 f(x)\mathrm{d}x = 3$, what is the value of $\int_1^2 2[f(x)+1]\mathrm{d}x$?

You must show all your working clearly. **[4 marks]**

4 Given that $(\sqrt[p]{x^{2p}})^q = x^p \times x^{q+1}$, find an expression for p in terms of q **[4 marks]**

5 A point P lies on the curve with equation $y = x^2 - 2$

The normal to the curve at P is parallel to the line with equation $x = -2y$

Find the equation of the tangent to the curve at P **[6 marks]**

6 a Describe the single geometrical transformation which maps $y=\sqrt{x}$ onto $y=3+\sqrt{x-4}$

[2 marks]

_____

_____

_____

_____

b i State the domain and range of the function $f(x)=3+\sqrt{x-4}$

[2 marks]

_____

_____

_____

_____

_____

_____

_____

ii Hence sketch the graph of $y=f(x)$ on the set of axes below.

[2 marks]

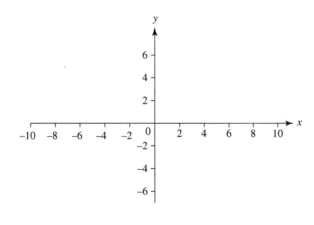

_____

_____

_____

_____

_____

6 c i $f(x)$ has an inverse function $f^{-1}(x)$

Find $f^{-1}(x)$ and state its domain and range.   **[6 marks]**

_____

_____

_____

_____

_____

_____

_____

_____

_____

_____

_____

_____

_____

_____

_____

ii Sketch the graph of $y = f^{-1}(x)$ on the set of axes below.   **[2 marks]**

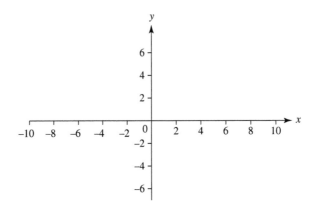

7 a Find the range of values of x for which the curve $f(x) = x^3 - 12x + 3$ is decreasing. **[3 marks]**

b Use the second derivative test to confirm that $f(x)$ has an inflection point at $(0, 3)$. **[3 marks]**

8 Given that θ is measured in radians, use the method of **differentiation by first principles** to find the derivative of the function $f(\theta) = \cos\theta$ with respect to θ

You should show all your working, but you may assume the small angle approximations for $\sin A$ and $\cos A$, and the compound angle formula for $\cos(A+B)$ **[6 marks]**

9 If $\dfrac{\mathrm{d}y}{\mathrm{d}x} = \dfrac{1}{x^2 - x}$, find an expression for y in terms of x

Give your answer as a **single** logarithmic term. [8 marks]

End of section A

75

10   Find the resultant of the vectors $4\mathbf{i} - 3\mathbf{j}$ and $7\mathbf{i} + \mathbf{j}$ **[1 mark]**

Circle your answer.

9          $11\mathbf{i} - 2\mathbf{j}$ $5 + \sqrt{50}$ $-3\mathbf{i} - 4\mathbf{j}$

11   A uniform rod AB of length 3 m and mass 6 kg is pivoted smoothly at A

The rod is held in equilibrium at an angle of 45° to the horizontal by a force **F** which acts perpendicular to the rod at B, as shown in the diagram.

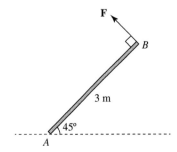

Calculate the magnitude of **F**   **[4 marks]**

_____

_____

_____

_____

_____

_____

_____

_____

_____

_____

_____

_____

12   A block of mass 12 kg rests on a surface which is inclined at 40° to the horizontal.

A light string is attached to the top corner of the block and is held taut at an angle of 30° above the slope. The string holds the block in limiting equilibrium so that the block is about to slide down the slope. The coefficient of friction between the block and the slope is 0.5

By first drawing a diagram to show all the forces acting on the block, find the tension in the string, **T**   **[7 marks]**

13   A ball is thrown from a window which is 12 m above horizontal ground.

The initial speed of the ball is 15 m s⁻¹ and it is thrown at an angle of 10° below the horizontal.

The ball strikes a vertical wall that lies a horizontal distance of 9 m away from the window, as shown in the diagram.

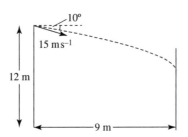

a   Calculate the height at which the ball strikes the wall.   **[5 marks]**

13 b   Calculate the speed of the ball and its direction of motion immediately before it hits the wall.   **[7 marks]**

14 A skydiver of mass 71 kg steps out of a hot-air balloon which is stationary and high above the ground.

Her velocity is v m s^{-1} at time t s after she leaves the balloon.

While falling, she experiences a resistive force (due to air resistance) of $5.8v$ N

a Draw a diagram to show the forces acting on the skydiver. **[2 marks]**

b The skydiver accelerates towards the ground until the resistive force is equal to her weight.

The velocity at which she travels when these two forces are equal is called her *terminal velocity*.

Calculate the terminal velocity of the skydiver. **[2 marks]**

c Use Newton's second law to show that $120 - v = 12.2\dfrac{dv}{dt}$

Give your working to 3 sf. **[3 marks]**

14 d   Find t when $v = 90\,\mathrm{m\,s^{-1}}$ **[8 marks]**

15 Particles P and Q, with masses 0.5 kg and 0.3 kg respectively, are attached to the ends of a light, inextensible string that passes over a smooth, light pulley.

The string is taut, and P rests in limiting equilibrium on a rough plane which is inclined at 55° to the horizontal, as shown in the diagram. P is on the point of slipping down the plane.

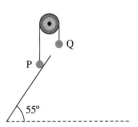

a Calculate the coefficient of friction, μ, between P and the plane. **[6 marks]**

15 b Another particle of mass 0.3 kg is attached to Q

The system is then released from rest.

Calculate the tension in the string and the acceleration of the particles. **[5 marks]**

End of questions

| 120 mins | **A Level Mathematics** | **AQA** |
| 100 marks | *Paper 3 (Set B)* | |

| Name | _____ | Class | _____ |
| Signature | _____ | Date | _____ |

Materials

You should have

- the booklet of formulae and statistical tables
- a graphical calculator.

Instructions

- Use a black pen for your working.
  Use a pencil for drawings.
- Answer **all** questions.
- Answer each question in the space provided for it; do **not** use the space provided for a different question. If you need extra space, ask for an additional answer book.
- All working should be inside the box drawn around each page.
- To avoid losing marks, show all necessary working.
- Include all rough working in this paper. If you do not want some work marked then cross it out.

Information

- Questions marks are shown in square brackets.
- There is a maximum of 100 marks available for this paper.

Advice

- Unless asked for a proof, you may quote any of the formulae in the booklet.
- You may not have to use all the answer space provided.

Question	Mark
1	
2	
3	
4	
5	
6	
7	
8	
9	
10	
11	
12	
13	
Total	

Section A

Answer **all** questions in the spaces provided.

1 Simplify fully the expression $4\left(\dfrac{x^2}{2}\right)^2$ **[1 mark]**

Circle your answer.

$$x^4 \qquad\qquad 2x^4 \qquad\qquad 4x^4 \qquad\qquad 16x^4$$

2 Work out $\dfrac{\mathrm{d}}{\mathrm{d}x}(\sin x^2)$ **[1 mark]**

Circle your answer(s).

$$2\sin x\cos x \qquad 2x\sin x\cos x \qquad \cos x^2 \qquad 2x\cos x^2$$

3   A temperature of 50° Fahrenheit is equal to 10° Celsius.

A temperature of 86° Fahrenheit is equal to 30° Celsius.

a   Find an equation relating Fahrenheit, F, to Celsius, C

Give your answer in the form $C = pF + q$, where p and q are rational numbers.   **[4 marks]**

b   Interpret, in context, the values of p and q **[2 marks]**

c   What temperature, in degrees Celsius, is exactly half the equivalent temperature in degrees Fahrenheit?   **[2 marks]**

4 a Find the **exact** solution to the inequality $\frac{1}{2}\left(\frac{3}{4}-\frac{x}{2}\right)-\frac{3}{8}\left(\frac{3x}{5}-\frac{1}{3}\right)<0$

You must show all your working. **[3 marks]**

b Find the **exact** solution to the inequality $2x^2-\frac{27x}{8}-\frac{11}{3}\le\frac{3x}{24}+\frac{1}{12}$

You must show all your working. **[5 marks]**

4 c Solve the inequality $1 < x^2 < 4$ **[3 marks]**

5 a Complete the rows in the table below.

Give each value correct to six decimal places. **[1 mark]**

x	2	3	5	6	8
y	7.68	6.144	3.932 16	3.145 728	2.013 265 92
$\log_{10}x$					
$\log_{10}y$					

b i Sketch the graph of $\log_{10} y$ against $\log_{10} x$ **[2 marks]**

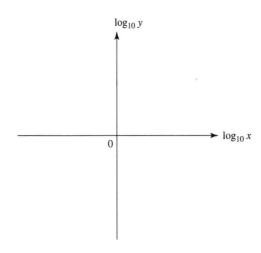

5   b   ii   Sketch the graph of $\log_{10} y$ against x **[2 marks]**

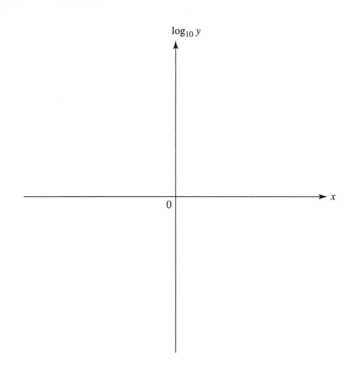

c   i   Hence determine whether the relationship between the data in the table is of the form $y = ax^b$ or of the form $y = ab^x$

Justify your answer.   **[3 marks]**

5 c ii Find the values of the constants a and b

Give the value of a correct to 2 significant figures, and the value of b correct to 1 significant figure. **[6 marks]**

6   Prove that $\dfrac{\sin^2 2x}{4} + \sin^4 x = 1 - \dfrac{\cot^2 x}{\operatorname{cosec}^2 x}$ **[6 marks]**

7 The curve defined by the parametric equations $x = t(t^2 - 4), y = 6(t^2 - 4)$ is sketched below.

a At what point does the curve cross itself?

Justify your answer. **[3 marks]**

_____

_____

_____

_____

_____

_____

_____

_____

_____

_____

_____

_____

_____

_____

_____

_____

_____

7 b Find the equations of both tangents to the curve at the point where the curve crosses itself. [6 marks]

End of section A

Section B

Answer **all** questions in the spaces provided.

8   Events A and B are not independent and are not mutually exclusive.

If $P(A) = 0.4$ and $P(B) = 0.7$, what is $P(A')$?   **[1 mark]**

Circle your answer.

| 0.1 | 0.28 | 0.5 | 0.6 |

9   The standard Normal density function describes a probability distribution and has a bell-shaped curve. It is usually written as $Z \sim N(0, 1)$

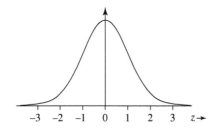

Use the fact that $P(Z < -0.4) = 0.3446$ and $P(Z < 0.2) = 0.5793$ to find the following to 3 significant figures without using your calculator.

You must show all your working.

a $P(Z > 0.2)$ **[2 marks]**

_____

_____

b $P(Z < 0.4)$ **[2 marks]**

_____

_____

c $P(-0.4 < Z < 0.2)$ **[2 marks]**

_____

_____

10   A bag contains a large number of balls. Each ball is either red, blue or green, and is either spotted or plain. The numbers of each type are given in this table.

	Red	Blue	Green
Spotted	34	41	75
Plain	27	23	50

a   A single ball is drawn from the bag at random. Show your working and calculate the probability that the drawn ball

    i   Is green and spotted,   **[1 mark]**

    ii   Is blue or plain,   **[1 mark]**

    iii   Is red, given that it is spotted.   **[2 marks]**

b   Are the events 'the ball is green' and 'the ball is plain' independent?
Explain your answer.   **[2 marks]**

11 The probability distribution for a random variable X is given by

$$P(X = x) = \frac{k}{1 + 2(x-2)^2} \quad \text{for } x = 0, 1, 2, 3, 4$$

where k is a rational number.

a In terms of k, find the probability of each value that X can take, and hence calculate the value of k **[4 marks]**

X is used to approximate a Normally distributed variable Y with mean 2 and standard deviation 0.75

b i Explain why $P(1.5 < Y < 2.5) \approx P(X = 2)$. **[1 mark]**

ii Calculate the probabilities of obtaining outcomes which round to each of 0, 1, 2, 3, and 4 with this Normal distribution. **[3 marks]**

11 c   Using your answers to **a** and **b**, comment on the validity of using X to approximate Y [2 marks]

_____

_____

_____

12   The following is an excerpt from a document detailing the sampling method for the Large Data Set.

'The Living Costs and Food Survey (LCF) sample for Great Britain is a multi-stage stratified random sample with clustering. It is drawn from the Small Users file of the Postcode Address File —the Post Office's list of addresses. The Northern Ireland sample is drawn as a random sample of addresses from the Land and Property Services Agency list.

The survey is a voluntary sample survey of private households run at household level. The survey is continuous, interviews being spread evenly over the year to ensure that seasonal effects are covered. Each household member over the age of seven keeps a diary of all their expenditure over a 2 week period. The diaries record expenditure and quantities of purchases of food and drink rather than consumption of food and drink.

In 2014 the survey collected the diaries of 121 250 people within 5144 households across the United Kingdom.'

a   Explain the meaning of the terms stratified sample and cluster sample.   [4 marks]

_____

_____

_____

_____

_____

_____

_____

_____

_____

_____

12 b Define the two sampling frames used in the survey. **[2 marks]**

c The survey is voluntary.

Give one reason why this might result in the data acquired being biased. **[1 mark]**

13 Bhushan is investigating the mass of fish eaten per person per week by a sample of 41 people who live in England. He uses data taken from the Large Data Set from 1974–2014 and presents his findings using the table and histogram shown below.

The raw data acquired by Bhushan (showing the mass of fish eaten per person per week by a sample of people who live in England) is shown in ascending order below.

118	122	123	128	129	131	137
140	140	141	141	144	144	144
144	144	146	146	146	146	147
147	147	147	148	148	148	148
149	150	151	155	156	157	158
158	158	161	165	167	170	

Mass of fish eaten per person per week (g)	$115 \leq m < 135$	$135 \leq m < 140$	$140 \leq m < 145$	$145 \leq m < 150$	$150 \leq m < 155$	$155 \leq m < 165$	$165 \leq m < 175$
Frequency			8	14	2	6	3

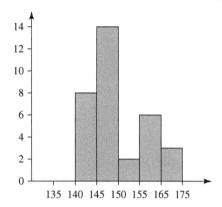

13 a i Label the axes of the histogram. [1 mark]

ii Use the raw data to fill in the missing values in the table. [2 marks]

iii Add the two missing bars to the histogram using the table of values. [2 marks]

b Alex says that the data collected by Bhushan looks like it could be drawn from a Normal distribution.

Give two features of Bhushan's data which support Alex's claim, and one feature of the data which might suggest that Alex's claim is wrong. [3 marks]

13 c Calculate the mean and variance of the sample. **[2 marks]**

d In 2014, the cost of fish was approximately 1.7 pence per gram.

If all the fish in Bhushan's sample had been purchased in 2014, use your answer to **c** to calculate the mean and variance in the cost of the sample. **[3 marks]**

13 e Bhushan collected the data to test the hypothesis that the average amounts of fish eaten per person per week could reasonably be 150 g. He assumes that each piece of raw data comes from a Normal distribution with variance equal to the sample variance in **c**

Perform Bhushan's hypothesis test at the 5% significance level, stating the hypotheses and conclusion clearly. **[7 marks]**

End of questions

| Name | _____ | Class | _____ |
| Signature | _____ | Date | _____ |

Materials

You should have

- the booklet of formulae and statistical tables
- a graphical calculator.

Instructions

- Use a black pen for your working.
  Use a pencil for drawings.
- Answer **all** questions.
- Answer each question in the space provided for it; do **not** use the space provided for a different question. If you need extra space, ask for an additional answer book.
- All working should be inside the box drawn around each page.
- To avoid losing marks, show all necessary working.
- Include all rough working in this paper. If you do not want some work marked then cross it out.

Information

- Questions marks are shown in square brackets.
- There is a maximum of 100 marks available for this paper.

Advice

- Unless asked for a proof, you may quote any of the formulae in the booklet.
- You may not have to use all the answer space provided.

Question	Mark
1	
2	
3	
4	
5	
6	
7	
8	
9	
10	
11	
12	
13	
Total	

Answer **all** questions in the spaces provided.

1 Which of the following is true when $\pi < \theta < \dfrac{3\pi}{2}$? **[1 mark]**

Circle your answer(s).

$\sin\theta < 0$ and $\cos\theta < 0$ $\sin\theta < 0$ and $\cos\theta > 0$

$\sin\theta > 0$ and $\cos\theta < 0$ $\sin\theta > 0$ and $\cos\theta > 0$

2 Which of the following is equal to $\sqrt{x^2 + y^2}$? **[1 mark]**

Circle your answer(s).

$x + y$ $(x+y)(x-y)$ $\pm(x+y)$ none of these

3 Which of the following is equal to $\ln(2e^x)$? **[1 mark]**

Circle your answer(s).

$x\ln 2$ $x\ln 2e$ $x + \ln 2$ $2x$

4   Simplify $\dfrac{1}{\dfrac{2}{\dfrac{3}{x}+1}+1}$ **[4 marks]**

5   **a**   Mark believes there is a solution to the following integration problem:

$$\int_{-2}^{3}\frac{1}{x^2}\,\mathrm{d}x$$

Sally disagrees and says that this problem has no solution.

Who is correct? Give two reasons to justify your answer.   **[2 marks]**

5 b To try to prove that he is correct, Mark offers the following solution to Sally:

$$\int_{-2}^{3} \frac{1}{x^2} \, dx = \left[-\frac{1}{x} \right]_{-2}^{3} = -\frac{1}{3} - \frac{1}{2} = -\frac{5}{6}$$

By considering the sign of $\frac{1}{x^2}$ and the sign of Mark's final answer, describe how

Sally might know immediately that Mark's solution must be incorrect. **[2 marks]**

6 a i Use the trapezium rule, with 8 strips, to estimate $\int_{2}^{4} f(x) \, dx$ where

$$f(x) = \frac{(x+3)^2}{x-1}, \ x \neq 1$$

Give your estimate to 3 decimal places. **[5 marks]**

6 a ii With the aid of an appropriate sketch, state whether your estimate in **6 a i** is likely to be an overestimate or an underestimate of the actual value.

Give a reason for your answer. **[3 marks]**

_____

_____

_____

_____

b i Show that $(x+3)^2$ can be written as $x(x-1)+a(x-1)+b$, where a an b are integers to be found. **[3 marks]**

_____

_____

_____

_____

_____

_____

_____

_____

_____

_____

_____

6 b ii Use your answer to part **6 b i**, along with a suitable method of integration, to find the **exact** value of $\int_{2}^{4} f(x)\,dx$

You must show every step of your working to gain full marks. **[5 marks]**

6 b iii Hence calculate, to the nearest 0.1%, the percentage error in your estimate of $\int_{2}^{4} f(x)\,dx$ which you found in **6 a i** by using the trapezium rule. **[2 marks]**

7 Given that θ is small and measured in radians, find **exact** approximations for the following expressions.

a $\dfrac{\cos 4\theta - 1}{\sin^{2} 3\theta}$ **[3 marks]**

7 **b** $\dfrac{4\cos 2\theta + \theta \sin^3 \theta}{\sin^4 \theta - 2\sin^2 2\theta + 4}$

[3 marks]

c $\dfrac{1 - \sin\left(\dfrac{\pi}{2} - \theta\right)}{\cos^2\left(\dfrac{\pi}{2} - \theta\right)}$

[3 marks]

8   a   Show that the circles $x^2 + y^2 + 8x - 6y - 11 = 0$ and $x^2 + y^2 - 16x - 16y + 79 = 0$ touch each other.   **[6 marks]**

8 b i Show that the two circles touch at the point with coordinates $\left(\dfrac{20}{13}, \dfrac{69}{13}\right)$ **[3 marks]**

ii Find the equation of the line which is a tangent to both circles at $\left(\dfrac{20}{13}, \dfrac{69}{13}\right)$

You must state any reasoning or assumptions that you use.   **[4 marks]**

9 The rate of growth of a population P at time t is directly proportional to the total population at time t days

a Write down a differential equation relating P and t **[1 mark]**

b By solving the differential equation, show that the general solution of the equation can be written in the form $P = Ae^{kt}$, where A and k are constants. **[3 marks]**

9 c   Initially, the population is 5.2 million. 14 days later, it has grown to 6 million.

Find the exact values of A and k

Find also the size of the population after a **further** 30 days.
Give your answer correct to 3 significant figures.   **[5 marks]**

10 The ellipse below is defined the parametric equations $x = 5\cos t - 2$, $y = 2\sin t - 1$, where t is a parameter.

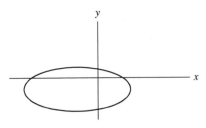

Find a Cartesian equation for the ellipse in the form $a(x+b)^2 + c(y+d)^2 = e$, where a, b, c, d and e are all positive integers. **[3 marks]**

11 a Use the substitution $u = \cos x$ to show that $\int \tan x \; \mathrm{d}x = \ln|\sec x| + c$, where c is an arbitrary constant.

Your working should be both clear and rigorous. **[6 marks]**

11 b Use a suitable trigonometric substitution to show that $\int \sec^4 x \tan x \; \mathrm{d}x = \dfrac{\sec^4 x}{4} + C$,

where C is an arbitrary constant. **[5 marks]**

11 c It is given that $\int \sec x \; dx = \ln|\sec x + \tan x| + K$, where K is an arbitrary constant.

Use this result, as well as your answers to parts **a** and **b**, to show that

$$\int \left[\sec x \tan x (\sec^3 x) - \sec x + \tan x \right] dx = \frac{\sec^4 x - 1}{4} + \ln\left(\frac{A \sec x}{\sec x + \tan x} \right)$$

where A is an arbitrary constant.

[4 marks]

12 a Show that $10x^2 - 11x + 1$ is a quadratic factor of the quartic expression $20x^4 - 72x^3 + 87x^2 - 38x + 3$ **[3 marks]**

b Hence write $20x^4 - 72x^3 + 87x^2 - 38x + 3$ as a product of four linear factors. **[2 marks]**

12 c You are given that $y = 20x^4 - 72x^3 + 87x^2 - 38x + 3$ has a turning point when $x \approx 0.35$ and another turning point when $x \approx 1.35$

Sketch the graph of y below, clearly indicating any points where your graph meets the coordinate axes. **[2 marks]**

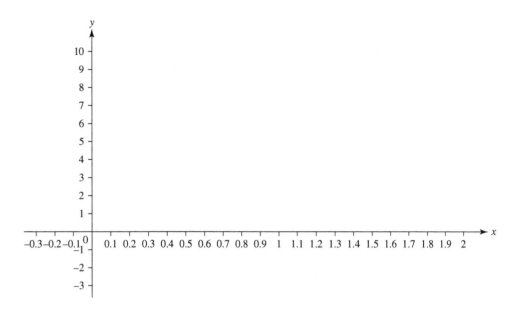

d Describe the geometrical transformation that will map y onto $20x^4 + 72x^3 + 87x^2 + 38x + 3$ **[1 mark]**

13 a i Show that $\dfrac{3}{n^2}\left[\dfrac{n(n+1)}{2}\right]+4=\dfrac{11}{2}+\dfrac{3}{2n}$

[2 marks]

ii Use the formula $S_n=\dfrac{n}{2}\left[2a+(n-1)d\right]$ to show that the arithmetic series given by

$$\sum_{i=1}^{n}i=1+2+3+\ldots+n \text{ has a sum of } S_n=\dfrac{n(n+1)}{2}$$

[2 marks]

13 b The definite integral $\int_{a}^{b}f(x)dx$, where $f(x)$ is a linear function, can be defined by

$$\int_{a}^{b}f(x)dx \;=\; \lim_{n\to\infty}\sum_{i=1}^{n}\left[\left(\dfrac{b-a}{n}\right)f(c_i)\right] \qquad \text{where } c_i=a+\left(\dfrac{b-a}{n}\right)i$$

i Verify that, for the definite integral $\int_{0}^{1}(3x+4)dx$, $c_i=\dfrac{i}{n}$

[1 mark]

13 b **ii** Use the Fundamental Theorem of Calculus to evaluate $\int_{0}^{1}(3x+4)\,\mathrm{d}x$ **[3 marks]**

---

iii Use the definition given at the beginning of the question to evaluate $\int_{0}^{1}(3x+4)\,\mathrm{d}x$

In your working, you may assume that $\sum_{i=1}^{n}\dfrac{i}{n} = \dfrac{1}{n}\sum_{i=1}^{n}i$ etc.   **[6 marks]**

End of questions

Name		Class	
Signature		Date	

Materials

You should have

- the booklet of formulae and statistical tables
- a graphical calculator.

Instructions

- Use a black pen for your working.
  Use a pencil for drawings.
- Answer **all** questions.
- Answer each question in the space provided for it; do **not** use the space provided for a different question. If you need extra space, ask for an additional answer book.
- All working should be inside the box drawn around each page.
- To avoid losing marks, show all necessary working.
- Include all rough working in this paper. If you do not want some work marked then cross it out.

Information

- Questions marks are shown in square brackets.
- There is a maximum of 100 marks available for this paper.

Advice

- Unless asked for a proof, you may quote any of the formulae in the booklet.
- You may not have to use all the answer space provided.

Question	Mark
1	
2	
3	
4	
5	
6	
7	
8	
9	
10	
11	
12	
13	
14	
Total	

Section A

Answer **all** questions in the spaces provided.

1 What is the range of validity of the binomial expansion of $\dfrac{1}{ax+b}$, where a and b are positive constants?   **[1 mark]**

Circle your answer.

$$|x| < \frac{b}{a} \qquad\qquad |x| < \frac{a}{b} \qquad\qquad |x| < a \qquad\qquad |x| < b$$

2 a Use the **product rule** to differentiate $\dfrac{e^{\frac{1}{2}x}}{2x}$ with respect to x **[3 marks]**

2 b Use a suitable **substitution** to integrate $\dfrac{1}{x \ln x}$ with respect to x **[4 marks]**

Use a suitable **substitution** to integrate $\dfrac{1}{x \ln x}$ with respect to x

3   The trapezium *ABCD* (shown below) has an area of $9(4\sqrt{3}+1)$ square units and a height of $2\sqrt{3}$ units.

The parallel sides *AD* and *BC* are such that $AD : BC = 2 : 1$

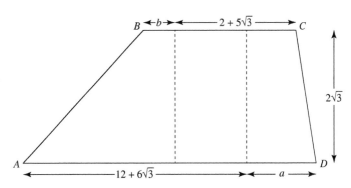

Find the exact lengths *a* and *b* [5 marks]

4   Prove that the value of the expression $2x^2 - 12x + 19$

is positive for all real values of x **[4 marks]**

5 **a**   The cubic equation $x^3 - x^2 + 3x - 4 = 0$ has a single root, α

Show that α lies between $x = 1$ and $x = 1.5$ **[3 marks]**

5 b Show that the cubic equation can be rearranged into the form $x = \sqrt{\dfrac{x^2 + 4}{x} - 3}$ [2 marks]

c Using the rearrangement given in part **b**, state a recurrence relation that you could use to find α

Using $x_1 = 1.25$, use your relation to find α correct to three decimal places.

You must show all your working. [5 marks]

5   d   The graph below shows the curve $y = \sqrt{\dfrac{x^2+4}{x}} - 3$ and the line $y = x$

On the graph, draw a cobweb or staircase diagram, with $x_1 = 1.25$, to show how convergence to α takes place via the recurrence relation in part **c**.

Indicate on your diagram the positions of x_2 and x_3 **[4 marks]**

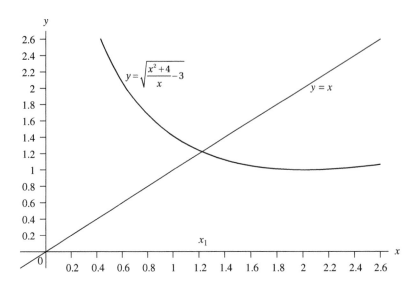

6 By implicitly differentiating both sides of $(x+\sin^2 y)^2 = (xy)^2$, find an expression for $\dfrac{dy}{dx}$

[5 marks]

7 **a** **i** What is the period, in radians, of the graph of $y = \cos bx$, where b is a positive constant? **[1 mark]**

ii What is the amplitude of the graph of $y = \cos bx$? **[1 mark]**

b **i** What is the period, in radians, of the graph of $y = a \cos bx$, where a and b are both positive constants? **[1 mark]**

ii What is the amplitude of the graph of $y = a \cos bx$? **[1 mark]**

c **i** What is the period, in radians, of the graph of $y = a \cos bx + c$, where a, b and c are all positive constants? **[1 mark]**

ii What is the amplitude of the graph of $y = a \cos bx + c$? **[1 mark]**

7 d The graph below shows the number of hours daylight experienced by a city over the course of a year.

The longest day occurs on June 21st, when the city experiences 17 hours of daylight.

The shortest day is December 21st, when there are only 7.5 hours of daylight.

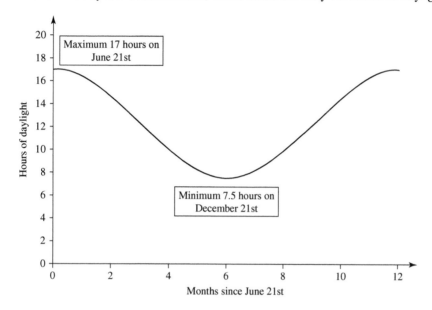

Write a trigonometric model (in radians) for the number of hours of daylight (*y*) in terms of the number of months since June 21st (*x*)

Give your model in the form $y = a\cos bx + c$, where *a*, *b* and *c* are positive constants. **[7 marks]**

7　d　　(cont.)

End of section A

Section B

Answer **all** questions in the spaces provided.

8   What feature of a velocity–time graph represents the displacement?   **[1 mark]**

Circle your answer.

    Area between graph and y-axis   Second derivative

    Gradient   Area between graph and x-axis

9   Relative to perpendicular axes x, y and z, points A and B have coordinates $(2, 6, -3)$ and $(5, 2, -3)$ respectively.

Find the unit vector in the direction of $\overrightarrow{AB}$

Give your answer in the form $a\mathbf{i} + b\mathbf{j} + c\mathbf{k}$, where $\mathbf{i}$, $\mathbf{j}$ and $\mathbf{k}$ are the unit vectors in the x-, y- and z-directions respectively.   **[4 marks]**

10   A uniform bar AB is 2.8 m long and weighs 80 N

Weights of 20 N and 40 N are attached to the bar at A and B respectively.

With the weights attached, the bar remains in horizontal equilibrium when supported at position C, which lies between A and B

a   Find the magnitude of the reaction at C **[2 marks]**

b   Find the distance from AC **[3 marks]**

11   **i** and **j** are unit vectors in the x- and y- directions respectively.

Initially, a particle is at the point with position vector $\mathbf{r}_0 = -12\mathbf{i} + 2\mathbf{j}$ m and is travelling with velocity $\mathbf{v}_0 = (8\mathbf{i} - 3\mathbf{j})\,\text{m s}^{-1}$

The particle accelerates uniformly for 8 seconds until it reaches the point with position vector $\mathbf{r} = 28\mathbf{i} + 10\mathbf{j}$ m

Find the acceleration of the particle.   **[6 marks]**

12 A force **F** acts on a particle of mass 6 kg, causing the particle to move in a straight line.

The particle's displacement, s, from the origin O at time t is found to be $s = t\mathrm{e}^{5-t}$ m

a Show that the velocity of the particle at time t will be given by $v = -\mathrm{e}^{5-t}(t-1)\,\mathrm{m\,s^{-1}}$ **[3 marks]**

b Find the magnitude of **F** when the particle is stationary. **[7 marks]**

12 b   (cont.)

13 A box of mass 8 kg is held at rest on a slope which is inclined at 35° to the horizontal.

The box is connected to a mass of 2 kg by a light, inextensible string which passes over a smooth pulley. The 2 kg mass hangs freely below the pulley, as shown in the diagram.

The coefficient of friction between the box and the slope is 0.3 and the system is initially held at rest in equilibrium.

When the system is released from rest, it takes 3 seconds for the box to slide s m to the bottom of the slope.

Find the distance s [9 marks]

13   **(cont.)**

14 a   A particle is projected from ground level on a horizontal surface.

The initial speed of the particle is u m s^{-1} and it is projected at an angle of $\theta°$ to the horizontal.

After t seconds have elapsed, the particle is at the point (x, y)

Show that the particle moves in a path given by the equation $y = x\tan\theta - \dfrac{g\sec^2\theta}{2u^2}x^2$ **[7 marks]**

14 b A waterskier performs a stunt where she takes off from a ramp at 9 m s^{-1} and tries to clear an obstacle that is 1m higher than the end of the ramp.

The obstacle lies in a vertical plane at a horizontal distance of 4 m from the end of the ramp.

Using the result from **a**, find the maximum and minimum angles that the ramp must make with the horizontal if the waterskier is to clear the obstacle.

You must show all your working. **[6 marks]**

c Another model of the waterskier takes account of air resistance acting on the water skier. How would you expect the maximum and minimum values of θ given by this model to be different to the values of θ you calculated in part **b**?

Give a reason for your answer. **[3 marks]**

End of questions

Name		Class	
Signature		Date	

Materials

You should have

- the booklet of formulae and statistical tables
- a graphical calculator.

Instructions

- Use a black pen for your working.
  Use a pencil for drawings.
- Answer **all** questions.
- Answer each question in the space provided for it; do **not** use the space provided for a different question. If you need extra space, ask for an additional answer book.
- All working should be inside the box drawn around each page.
- To avoid losing marks, show all necessary working.
- Include all rough working in this paper. If you do not want some work marked then cross it out.

Information

- Questions marks are shown in square brackets.
- There is a maximum of 100 marks available for this paper.

Advice

- Unless asked for a proof, you may quote any of the formulae in the booklet.
- You may not have to use all the answer space provided.

Question	Mark
1	
2	
3	
4	
5	
6	
7	
8	
9	
10	
11	
12	
13	
14	
15	
16	
17	
Total	

Section A

Answer **all** questions in the spaces provided.

1   What is the derivative of 2^x with respect to x?   **[1 mark]**

Circle your answer(s).

$$2x^{x-1} \qquad 2^{x-1}x \qquad 2^x \ln 2 \qquad x \ln 2$$

2 $\dfrac{\mathrm{d}x}{\mathrm{d}t} = kx$, where k is a constant

Which of these are general solutions to the above differential equation?
(A and c are arbitrary constants.)   **[1 mark]**

Circle your answer(s).

$$x = \mathrm{e}^{kt+c} \qquad x = A\mathrm{e}^{kt} \qquad x = kt + c \qquad x = \frac{kx^2}{2} + c$$

3   If $x = \cos^2 y$, find $\dfrac{dy}{dx}$ in terms of a trigonometric ratio of $2y$ **[5 marks]**

4   Blaise solves the inequality $x^2 < a^2$ for x, where a is a real number.

He writes his solution as

$$\{x : -a < x\} \cap \{x : x < a\}$$

Is Blaise correct? Explain your answer.   **[2 marks]**

5 a   On the same set of axes below, sketch the graphs of $y = |4x - 1|$ and $y = |3 - x|$

Label clearly any points where the graphs meet the coordinate axes.   **[4 marks]**

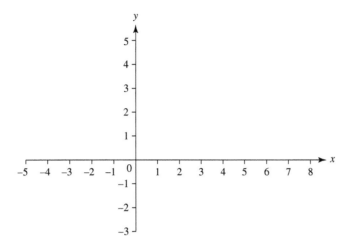

5 b Find the exact values of the two pairs of coordinates where the graphs of $y = |4x - 1|$ and $y = |3 - x|$ intersect. **[6 marks]**

c Hence, write down the exact solution to the inequality $|3 - x| > |4x - 1|$ **[1 mark]**

6   The denominator of a fraction is five more than its numerator.

Both numerator and denominator are positive.

When four is added to both the numerator and denominator, the value of the fraction increases by $\dfrac{5}{24}$

What is the original fraction?

You must show all your working.   **[6 marks]**

7   Vince chooses any two integers between 0 and 9   For example, 4 and 5

He forms a two-digit number using these two integers.   45

He then reverses the digits to form a second two-digit number.   54

Vince observes that the sum of both two-digit numbers
seems always to be a multiple of 11 $45 + 54 = 99 = 9 \times 11$

Prove that Vince's observation will always be true for **any** two-digit number.   **[4 marks]**

8   Describe the single geometrical transformation which maps

$f(x) = \cos 2x$ onto $g(x) = \cos\left(2x + \dfrac{\pi}{6}\right)$ **[2 marks]**

9 a Write down the three terms (**not** the sum) of the geometric sequence represented

by $\sum_{r=1}^{3} 2\left(\dfrac{\sqrt{2}}{2}\right)^{r-1}$

[1 mark]

b Find $\sum_{r=10}^{\infty} 2\left(\dfrac{\sqrt{2}}{2}\right)^{r-1}$

Write your answer in the form $a+b\sqrt{2}$, where a and b are rational numbers.

You must show all your working.

[6 marks]

10 a   It is given that $\sin(A-B) \equiv \sin A \cos B - \cos A \sin B$

By replacing A with $\dfrac{\pi}{2} - A$, prove that $\cos(A+B) \equiv \cos A \cos B - \sin A \sin B$ **[3 marks]**

b   Use the result in part **a** to deduce that $\cos^2 A \equiv \dfrac{1 + \cos 2A}{2}$ **[3 marks]**

10 c Hence find $\int \cos^4 A \, dA$ **[5 marks]**

End of section A

Section B

Answer **all** questions in the spaces provided.

11 Events A and B are **independent** events.

Which of the following is **not** true about A and B? **[1 mark]**

Circle your answer.

$P(B|A) = P(A)$ $\qquad\qquad\qquad\qquad$ $P(A \cap B) = P(A) \times P(B)$

$P(A|B) = P(A)$ $\qquad\qquad\qquad\qquad$ $P(A \cup B) = P(A) + P(B) - P(A \cap B)$

12 A test is performed to see if the mean, μ, of a variable which follows a Normal distribution has increased from its previous value, which was 7

Which of the following is the alternative hypothesis for the test? **[1 mark]**

Circle your answer.

$\text{H}_1 : \mu < 7$ $\qquad$ $\text{H}_1 : \mu \neq 7$ $\qquad$ $\text{H}_1 : \mu \geq 7$ $\qquad$ $\text{H}_1 : \mu > 7$

13 A group of 16 students each take both a spelling test and a mental arithmetic test. Each test is marked out of a total of 10 marks. For the 16 students, the scores of both tests are presented in this scatter graph.

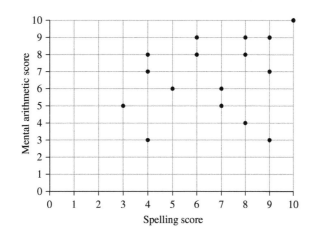

a Describe the correlation between the spelling score and mental arithmetic score. **[1 mark]**

_____

_____

b The correlation coefficient for the data is 0.268 4

Use this data to perform a hypothesis test at the 5% significance level to decide if there is any correlation between students' scores in the spelling test and the mental arithmetic test.

The critical values are ±0.514 **[5 marks]**

_____

_____

_____

_____

_____

_____

_____

_____

_____

14 When an unbiased, six-sided die is rolled, the outcome is a face showing a number between 1 and 6

The events $E = \{2,4,6\}$ and $O = \{1,3,5\}$ are exhaustive.

a State what is meant by the term exhaustive. **[1 mark]**

The dice is rolled 4 times. The event $EEEO$ shows that the first three results were in the set E and the fourth result is in the set O

b List all 16 possible outcomes from the 4 rolls. **[1 mark]**

c State the probability that a random outcome contains at least three results from set E **[1 mark]**

d Find the probability that a random outcome contains at least three results from set E, given that it contains at least two results from set E **[1 mark]**

15 A polling company wants to see if a newly-announced policy by a parliamentary party is popular among a constituency. They assume that one third of all people will agree with the policy and the rest will disagree with it.

The company takes a sample of size 30 by inviting people from the constituency to attend a lunchtime meeting in exchange for a small monetary gift.

a State one potential weakness of this sampling method. **[1 mark]**

b Let the random variable X represent the number of people in the sample who agree with the policy.

State the distribution of X **[1 mark]**

c State the mean and variance of X **[2 marks]**

d Using your distribution in **b**, find the probability that at most 5 people in the sample agree with the policy. **[1 mark]**

15 e After the meeting, it turned out that half the group agreed with the policy.

Assuming that the group forms a random sample, perform a hypothesis test at the 5% level to decide if the company's assumption about the level of the support for the policy was an underestimate. **[7 marks]**

16 Shaan is investigating the amount of tea drunk by English people over time.
He collects data from 1974–2014 and groups it in the following table.
Each data item is the average amount (in g) consumed per person per week over
that year.

Mass range (m) (g)	Cumulative Frequency
$0 < m \le 25$	0
$25 < m \le 27$	3
$27 < m \le 30$	8
$30 < m \le 35$	15
$35 < m \le 40$	19
$40 < m \le 45$	24
$45 < m \le 55$	31
$55 < m \le 65$	38
$65 < m \le 68$	41

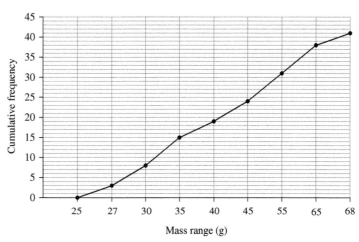

a Estimate the median.   **[1 mark]**

b Estimate the interquartile range (IQR).   **[3 marks]**

c Estimate the probability that for a randomly-chosen year the average amount of tea
consumed per person per week exceeds 50 g.   **[2 marks]**

16 d Data is also collected over the same period of the amount of coffee (in g) consumed per person per week on average each year.

Mass range	Cumulative Frequency
$0 < m \le 13$	0
$13 < m \le 15$	1
$15 < m \le 16$	6
$16 < m \le 17$	12
$17 < m \le 18$	17
$18 < m \le 19$	24
$19 < m \le 20$	33
$20 < m \le 21$	39
$21 < m \le 22$	41

Draw a cumulative frequency diagram for this data. **[3 marks]**

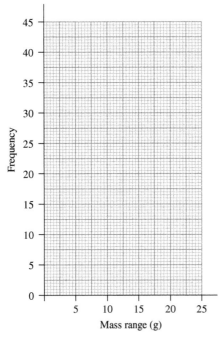

e Compare the two distributions, referring to the central tendency and the spread. **[2 marks]**

17 Iris is studying the amount of white bread purchased per person per week across regions of England.

She collects data over a period of 14 years for five different regions of England, and generates box-and-whisker plots from the summary statistics given in this table.

	Min	LQ	Med	UQ	Max
England	216	260	297	336	439
North East	238	289	321	357	452
East Midlands	298	359	390	418	542
South West	196	257	292	325	426
London	137	196	217	262	366

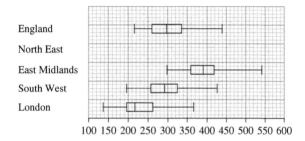

a Add to the diagram the box-and-whisker plot for the North East. **[3 marks]**

b State the region with the largest range and the region with the smallest range. **[2 marks]**

17 c Verify that the East Midlands contains at least one outlier.

Justify whether or not the outlier(s) should be excluded from the data. **[4 marks]**

d Cluster sampling is to be used for further studies.

From which single region would it be best to take a sample to get a reflection of the whole of England? Justify your choice. **[2 marks]**

e State one advantage and one disadvantage of taking a sample in this way. **[2 marks]**

f Define the term simple random sampling and explain why cluster sampling does not provide a simple random sample. **[2 marks]**

End of questions

Mathematical formulae
For A Level Maths

Pure Mathematics

Binomial series

$$(a+b)^n = a^n + \binom{n}{1}a^{n-1}b + \binom{n}{2}a^{n-2}b^2 + \cdots + \binom{n}{r}a^{n-r}b^r + \cdots + b^n \qquad (n \in \mathbb{N})$$

where $\displaystyle \binom{n}{r} = {}^nC_r = \frac{n!}{r!(n-r)!}$

$$(1+x)^n = 1 + nx + \frac{n(n-1)}{1.2}x^2 + \ldots + \frac{n(n-1)\ldots(n-r+1)}{1.2\ldots r}x^r + \ldots \qquad (|x|<1, \ n \in \mathbb{Q})$$

Arithmetic series

$$S_n = \frac{1}{2}n(a+l) = \frac{1}{2}n[2a+(n-1)d]$$

Geometric series

$$S_n = \frac{a(1-r^n)}{1-r} \qquad\qquad S_\infty = \frac{a}{1-r} \textit{ for } |r|<1$$

Trigonometry: small angles

For small angle θ, measured in radians:

$$\sin\theta \approx \theta \qquad\qquad \cos\theta \approx 1 - \frac{\theta^2}{2} \qquad \tan\theta \approx \theta$$

Trigonometric identities

$$\sin(A \pm B) = \sin A \cos B \pm \cos A \sin B$$

$$\cos(A \pm B) = \cos A \cos B \mp \sin A \sin B$$

$$\tan(A \pm B) = \frac{\tan A \pm \tan B}{1 \mp \tan A \tan B} \qquad \left(A \pm B \neq \left(k+\frac{1}{2}\right)\pi\right)$$

Differentiation

$f(x)$	$f'(x)$
$\tan x$	$\sec^2 x$
$\operatorname{cosec} x$	$-\operatorname{cosec} x \cot x$
$\sec x$	$\sec x \tan x$
$\cot x$	$-\operatorname{cosec}^2 x$
$\dfrac{f(x)}{g(x)}$	$\dfrac{f'(x)g(x)-f(x)g'(x)}{(g(x))^2}$

Differentiation from first principles

$$f'(x) = \lim_{h \to 0} \frac{f(x+h)-f(x)}{h}$$

Integration

$$\int u \frac{dv}{dx}\, dx = uv - \int v \frac{du}{dx}\, dx$$

$$\frac{f'(x)}{f(x)} = \ln|f(x)| + c$$

$f(x)$	$\int f(x)\,dx$
$\tan x$	$\ln|\sec x| + c$
$\cot x$	$\ln|\sin x| + c$

Numerical solution of equations

The Newton–Raphson iteration for solving $f(x)=0$: $x_{n+1} = x_n - \dfrac{f(x_n)}{f'(x_n)}$

Numerical integration

The trapezium rule: $\displaystyle\int_a^b y\,dx \approx \frac{1}{2}h\{(y_0 + y_n) + 2(y_1 + y_2 + \cdots + y_{n-1})\}$, where $h = \dfrac{b-a}{n}$

Mechanics

Constant acceleration

$$s = ut + \frac{1}{2}at^2 \qquad \mathbf{s} = \mathbf{u}t + \frac{1}{2}\mathbf{a}t^2$$

$$s = vt - \frac{1}{2}at^2 \qquad \mathbf{s} = \mathbf{v}t - \frac{1}{2}\mathbf{a}t^2$$

$$v = u + at \qquad \mathbf{v} = \mathbf{u} + \mathbf{a}t$$

$$s = \frac{1}{2}(u+v)t \qquad \mathbf{s} = \frac{1}{2}(\mathbf{u} + \mathbf{v})t$$

$$v^2 = u^2 + 2as$$

Probability and statistics

Probability

$P(A \cup B) = P(A) + P(B) - P(A \cap B)$

$P(A \cap B) = P(A) \times P(B|A)$

Discrete distributions

Distribution of X	$P(X = x)$	Mean	Variance
Binomial $B(n, p)$	$\binom{n}{x} p^x (1-p)^x$	np	$np(1 - p)$

Sampling distributions

For a random sample of n observations from $N(\mu, \sigma^2)$

$$\frac{\bar{X} - \mu}{\frac{\sigma}{\sqrt{n}}} \sim N(0, 1)$$

Statistical tables
For A Level Maths

The following statistical table will be provided for you.

Critical values of the product moment correlation coefficient

The table gives the critical values, for different significance levels, of the product moment correlation coefficient, r, for varying sample sizes, n

| One tail | 10% | 5% | 2.5% | 1% | 0.5% | One tail |
Two tail	20%	10%	5%	2%	1%	Two tail
n						n
4	0.8000	0.9000	0.9500	0.9800	0.9900	4
5	0.6870	0.8054	0.8783	0.9343	0.9587	5
6	0.6084	0.7293	0.8114	0.8822	0.9172	6
7	0.5509	0.6694	0.7545	0.8329	0.8745	7
8	0.5067	0.6215	0.7067	0.7887	0.8343	8
9	0.4716	0.5822	0.6664	0.7498	0.7977	9
10	0.4428	0.5494	0.6319	0.7155	0.7646	10
11	0.4187	0.5214	0.6021	0.6851	0.7348	11
12	0.3981	0.4973	0.5760	0.6581	0.7079	12
13	0.3802	0.4762	0.5529	0.6339	0.6835	13
14	0.3646	0.4575	0.5324	0.6120	0.6614	14
15	0.3507	0.4409	0.5140	0.5923	0.6411	15
16	0.3383	0.4259	0.4973	0.5742	0.6226	16
17	0.3271	0.4124	0.4821	0.5577	0.6055	17
18	0.3170	0.4000	0.4683	0.5425	0.5897	18
19	0.3077	0.3887	0.4555	0.5285	0.5751	19
20	0.2992	0.3783	0.4438	0.5155	0.5614	20
21	0.2914	0.3687	0.4329	0.5034	0.5487	21
22	0.2841	0.3598	0.4227	0.4921	0.5368	22
23	0.2774	0.3515	0.4132	0.4815	0.5256	23
24	0.2711	0.3438	0.4044	0.4716	0.5151	24
25	0.2653	0.3365	0.3961	0.4622	0.5052	25
26	0.2598	0.3297	0.3882	0.4534	0.4958	26
27	0.2546	0.3233	0.3809	0.4451	0.4869	27
28	0.2497	0.3172	0.3739	0.4372	0.4785	28
29	0.2451	0.3115	0.3673	0.4297	0.4705	29
30	0.2407	0.3061	0.3610	0.4226	0.4629	30
31	0.2366	0.3009	0.3550	0.4158	0.4556	31
32	0.2327	0.2960	0.3494	0.4093	0.4487	32
33	0.2289	0.2913	0.3440	0.4032	0.4421	33

34	0.2254	0.2869	0.3388	0.3972	0.4357	**34**
35	0.2220	0.2826	0.3338	0.3916	0.4296	**35**
36	0.2187	0.2785	0.3291	0.3862	0.4238	**36**
37	0.2156	0.2746	0.3246	0.3810	0.4182	**37**
38	0.2126	0.2709	0.3202	0.3760	0.4128	**38**
39	0.2097	0.2673	0.3160	0.3712	0.4076	**39**
40	0.2070	0.2638	0.3120	0.3665	0.4026	**40**
41	0.2043	0.2605	0.3081	0.3621	0.3978	**41**
42	0.2018	0.2573	0.3044	0.3578	0.3932	**42**
43	0.1993	0.2542	0.3008	0.3536	0.3887	**43**
44	0.1970	0.2512	0.2973	0.3496	0.3843	**44**
45	0.1947	0.2483	0.2940	0.3457	0.3801	**45**
46	0.1925	0.2455	0.2907	0.3420	0.3761	**46**
47	0.1903	0.2429	0.2876	0.3384	0.3721	**47**
48	0.1883	0.2403	0.2845	0.3348	0.3683	**48**
49	0.1863	0.2377	0.2816	0.3314	0.3646	**49**
50	0.1843	0.2353	0.2787	0.3281	0.3610	**50**
60	0.1678	0.2144	0.2542	0.2997	0.3301	**60**
70	0.1550	0.1982	0.2352	0.2776	0.3060	**70**
80	0.1448	0.1852	0.2199	0.2597	0.2864	**80**
90	0.1364	0.1745	0.2072	0.2449	0.2702	**90**
100	0.1292	0.1654	0.1966	0.2324	0.2565	**100**

Objectives checklist

Use these checklists to assess your confidence with the topics and A Level Maths.

Ch	Objective	MyMaths	InvisiPen	No	Almost	Yes!
12 Algebra 2	Make logical deductions and prove statements directly by exhaustion, by counter example and by contradiction.	2254	12S1A	☐	☐	☐
	Understand and use functions.	2049, 2135, 2138, 2139, 2142, 2261	12S2B	☐	☐	☐
	Understand and use parametric equations.	2224, 2262	12S3A	☐	☐	☐
	Understand and use algebraic fractions in all their forms.	2200, 2259	12S4B	☐	☐	☐
	Decompose fractions into partial fractions.	2260	12S5A	☐	☐	☐
13 Sequences	Use the binomial expansion and recognise the range of validity.	2204, 2205	13S1A	☐	☐	☐
	Use the binomial expansion to estimate the value of a surd.	2204, 2205	–	☐	☐	☐
	Understand if a sequence is increasing or decreasing.	2264	13S2B	☐	☐	☐
	Work out the order of a periodic sequence.	2264	–	☐	☐	☐
	Work out the nth term and the sum of an arithmetic series.	2039	13S3A	☐	☐	☐
	Work out the nth term and the sum of a geometric series.	2040	13S4B	☐	☐	☐
	Evaluate a series given in sigma notation.	2040	–	☐	☐	☐
14 Trigonometric identities	Convert between degrees and radians and use radians in problems.	2050, 2266	14S1A	☐	☐	☐
	Use reciprocal and inverse trigonometric functions.	2155, 2156	14S2B	☐	☐	☐
	Use trigonometric formulae for compound angles, double angles and half angles.	2157, 2158, 2262	14S3A	☐	☐	☐
	Find and use equivalent forms for $a\cos\theta + b\sin\theta$	2159	14S4B	☐	☐	☐
	Solve equations using trigonometric formulae to simplify expressions.	2159	14S4B	☐	☐	☐
15 Differentiation 2	Find points of inflection and determine when a curve is convex or concave.	2271	15S1A	☐	☐	☐
	Use $\lim\limits_{\theta\to 0}\dfrac{\sin\theta}{\theta}=1$ and $\lim\limits_{\theta\to 0}\dfrac{1-\cos\theta}{\theta}=0$	2165	–	☐	☐	☐
	Differentiate $\sin x$, $\cos x$, e^x, a^x and $\ln x$	2165, 2161	15S2B, 15S3A	☐	☐	☐
	Use the product and quotient rule for differentiation.	2163, 2164	15S4B	☐	☐	☐
	Use the chain rule for differentiation.	2162, 2166	15S5A	☐	☐	☐
	Find the derivative of a function that is defined implicitly.	2223	15S7A	☐	☐	☐
	Find the derivative of an inverse function.	2272	15S6B	☐	☐	☐
	Find the derivative of a function that is defined parenthetically.	2222	15S8B	☐	☐	☐

Objectives checklist

Ch	Objective	MyMaths	InvisiPen	No	Almost	Yes!
16 Integration and differential equations	Integrate a set of standard functions, f(x) and the related functions, f($ax + b$)	2057, 2168, 2170, 2218	16SA	☐	☐	☐
	Find the area between two curves.	2274	–	☐	☐	☐
	Simplify an integral by changing the variable, referred to as *substitution*.	2167, 2169, 2216, 2219	16S2B	☐	☐	☐
	Use integration by parts to integrate the product of two functions.	2171, 2220	16S3A	☐	☐	☐
	Simplify an integral by decomposing a rational function into partial fractions.	2217	16S4B	☐	☐	☐
	Understand the meaning of the expression 'differential equation'.	2226, 2227	16S5A	☐	☐	☐
	Use integration where the variables are separable.	2226, 2227	16S5A	☐	☐	☐
17 Numerical Methods	Use the change of sign method to find and estimate the root(s) of an equation.	2173	17S1B	☐	☐	☐
	Use an iterative formula to estimate the root of an equation.	2174	17S2A	☐	☐	☐
	Recognise the conditions that cause an iterative sequence to converge.	2174	–	☐	☐	☐
	Use the Newton–Raphson method to estimate the root of an equation.	2176	17S3B	☐	☐	☐
	Use the trapezium rule to find the area under a curve.	2060	17S4A	☐	☐	☐
18 Motion in two dimensions	Use the constant acceleration equations for motion in two dimensions.	2290	18S1B	☐	☐	☐
	Use calculus to solve problems in two-dimensional motion with variable acceleration.	2291	18S2A	☐	☐	☐
	Solve problems involving the motion of a projectile under gravity.	2198, 2199	18S3B	☐	☐	☐
	Analyse the motion of an object in two dimensions under the action of a system of forces.	2192	18S4A	☐	☐	☐
19 Forces 2	Manipulate vectors in three dimensions, and solve geometrical problems.	2208	–	☐	☐	☐
	Understand that there is a maximum value that the frictional force can take (μR) and that it takes this value when the object is moving or on the point of moving.	2193	–	☐	☐	☐
	Resolve in suitable directions to find unknown forces when the system is at rest or has constant acceleration.	2190, 2191	19S2A	☐	☐	☐
	Use constant acceleration formulae for problems involving blocks on slopes or blocks connected by pulleys.	2191	19S1B	☐	☐	☐
	Solve differential equations which arise from problems involving $F = ma$	2194	–	☐	☐	☐
	Take moments about suitable points and resolve in suitable directions to find unknown forces.	2197	19S3B	☐	☐	☐

Objectives checklist

Ch	Objective	MyMaths	InvisiPen	No	Almost	Yes!
20 Probability and continuous random variables	Calculate conditional probabilities from data given in different forms.	2092, 2095	20S1A	☐	☐	☐
	Apply binomial and Normal probability models in different circumstances.	2113, 2120, 2121, 2292	20S2B, 20S3A	☐	☐	☐
	Use data to assess the validity of probability models.	2286	–	☐	☐	☐
	Solve problems involving both binomial and Normal distributions.	2286	20S4B	☐	☐	☐
21 Hypothesis testing 2	State null and alternative hypotheses when testing for correlation.	2287	21S1A	☐	☐	☐
	Compare a given PMCC to a critical value or its p-value to the significance level, and use this comparison to decide whether to accept or reject the null hypothesis.	2287	21S1A	☐	☐	☐
	Decide what the conclusion means in context about the correlation.	2287	21S1A	☐	☐	☐
	State null and alternative hypotheses when testing the mean of a Normal distribution.	2288	21S2B	☐	☐	☐
	Calculate the test statistic, compare it to a critical value or compare its p-value to the significance level, and use this comparison to decide whether to accept or reject the null hypothesis.	2288	21S2B	☐	☐	☐
	Decide what the conclusion means in context about the mean of the distribution.	2288	21S2B	☐	☐	☐

Answers

Paper 1 (Set A)

1 $\dfrac{3\cos x}{\cos x(\sin x \cos x - \cos^2 x)} = \dfrac{3}{\cos x(\sin x - \cos x)}$

2 $2xe^{x^2}$

3 $x\ln 2x - x + c$

4 **a** 6 cm **b** $(6\pi - 9\sqrt{3})\text{ cm}^2$

5 **a** **i** Gradient of 1st line = 2. Gradient of 2nd line = 2.

Gradient of 3rd line = −2. Gradient of 4th line = −2.

ii *ABCD* is a parallelogram, because it has two pairs of parallel sides.

b **i** Solve simultaneously each pair of non–parallel equations.

ii $\left(-\dfrac{1}{8}, \dfrac{3}{4}\right)$, $\left(\dfrac{5}{4}, \dfrac{7}{2}\right)$, $\left(\dfrac{21}{8}, -\dfrac{19}{4}\right)$, $(4, -2)$

c 3.07 cm (3 sf), 6.15 cm (3 sf)

6 **a** Assume that the contrary is true i.e. $\sqrt{2}$ is rational.

Then there exist integers p and q (with no common factors) such that

$\sqrt{2} = \dfrac{p}{q}$

$\Rightarrow 2 = \dfrac{p^2}{q^2} \Rightarrow p^2 = 2q^2$

$\Rightarrow p^2$ is even $\Rightarrow p$ is even

If p is even, then it is a multiple of 2

$\Rightarrow p = 2n$, where n is an integer.

Also, $\sqrt{2} = \dfrac{p}{q} \Rightarrow q^2 = \dfrac{p^2}{2}$

$\Rightarrow q^2 = \dfrac{(2n)^2}{2} = 2n^2$

$\Rightarrow q^2$ is even $\Rightarrow q$ is even.

But if p and q are both even, then they have a common factor of 2

This **contradicts** our original assumption.

$\sqrt{2}$ is **not** rational.

$\therefore \sqrt{2}$ is irrational.

b $\sqrt[8]{2}$

c **i** $4\sqrt{2}$ **ii** $16 + 8\sqrt{2}$

7 **a** $1 < k < 4$ **b** **i** 6 **ii** $3x + 4$

8 **a** $x = -7, x = 2$ **b** $y = \dfrac{1}{2}x + 15$ **c** $\dfrac{119}{4} + 14\ln 2$

9 **a** $\dfrac{\cos^2\theta}{\cos^2\theta} + \dfrac{\sin^2\theta}{\cos^2\theta} \equiv \dfrac{1}{\cos^2\theta}$

$\Rightarrow 1 + \tan^2\theta \equiv \sec^2\theta$

b $\theta = -0.905$, $\theta = 0.905$, $\theta = 5.38$ to 3 sf

c $x = -2.26$, $x = 2.26$, $x = 4.02$ to 3 sf

10 **a** Let $f(x) = x^2 - 6$

Then $f(2.4) = -0.24 < 0$

and $f(2.5) = 0.25 > 0$

$f(x)$ is **continuous** on the interval

(0.24, 0.25) and there is a **change of sign**.

$\therefore f(x) = 0$ has a root between $x = 2.4$ and $x = 2.5$

b 2.45

c 1.320469

d No, her procedure will not succeed as the gradient of the curve at x_0 is zero (or close to zero).

11 **a** $\sec^2 y$ **b** $\dfrac{1}{1 + x^2}$

Paper 2 (Set A)

1 $\ln 5$

2 Stretch parallel to the x-axis **and** stretch parallel to the y-axis

3 $\sqrt{\dfrac{11}{3}}$, $(-1, 2)$

4 $\dfrac{3}{4x} + \dfrac{2}{x^2} + \dfrac{1}{x-5} - \dfrac{7}{4(x-4)}$

5 **a** **i**

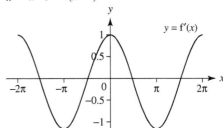

ii $y = \cos x$

b **i**

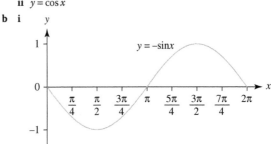

ii $y = -\sin x$

6 **a** π

b Both radius (or radius squared) and area cannot be negative.

c **i** $y = \sqrt{r^2 - x^2}$

$\text{Area required} = \displaystyle\int_0^r \sqrt{r^2 - x^2}\, dx$

Using $x = r\sin\theta$,

$x = 0 \Rightarrow \theta = \sin^{-1}\left(\dfrac{0}{r}\right) = 0$

$x = r \Rightarrow \theta = \sin^{-1}\left(\dfrac{r}{r}\right) = \dfrac{\pi}{2}$

Then area required becomes

$\displaystyle\int_0^{\frac{\pi}{2}} \sqrt{r^2 - r^2\sin^2\theta}\, \dfrac{dx}{d\theta}\, d\theta$

$= r\displaystyle\int_0^{\frac{\pi}{2}} \sqrt{1 - \sin^2\theta}\,(r\cos\theta)\, d\theta$

$= r^2\displaystyle\int_0^{\frac{\pi}{2}} \cos^2\theta\, d\theta$

ii $\dfrac{1}{4}\pi r^2$

d πr^2

7 **a** -1260 **b** $\dfrac{1}{8}$ **c** $\dfrac{3925}{16}$

8 a, b, c

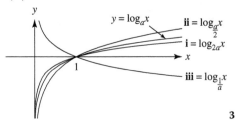

3

9 196 N

10 30.9 Nm (3 sf) clockwise

11 a i Gradient of graph = acceleration of the particle

    ii $a = \dfrac{\Delta v}{\Delta t}$

        $a = \dfrac{v-u}{t}$

        $at = v-u$

        $v = u + at$

  b 20.3 m

12 a Perpendicular to the wall

    Since their wall is modelled as smooth, there is no friction to cause a component of the reaction in the vertical direction.

  b 35.7 N

13 2.13 m s^{-2}

14 a 41°   **b** 7.8 s (2 sf)

15 a $F = \mu R$ N

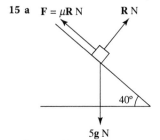

  b 37.5 N

  c 4.05 m s^{-2}

16 a $\dfrac{20\pi}{6}\cos\left(\dfrac{\pi}{6}t\right)\mathbf{i} - 3t^{-\frac{1}{2}}\mathbf{j}$

  b i $\dfrac{100\pi}{6}\cos\left(\dfrac{\pi}{6}t\right)\mathbf{i} - 15t^{-\frac{1}{2}}\mathbf{j}$

    ii $t = 9 \Rightarrow \mathbf{F} = \dfrac{100\pi}{6}\cos\left(\dfrac{9\pi}{6}\right)\mathbf{i} - 15\times 9^{-\frac{1}{2}}\mathbf{j}$

        $= 0\mathbf{i} - 5\mathbf{j}$

    Hence $|\mathbf{F}| = 5$

    Force acts in the negative **j**-direction.

  c $\mathbf{r} = \left(100 - \dfrac{120}{\pi}\cos\left(\dfrac{\pi}{6}t\right)\right)\mathbf{i} + \left(208 - 4t^{\frac{3}{2}}\right)\mathbf{j}$

Paper 3 (Set A)

1 n

2 $(x^4+1)(x^2+1)(x+1)(x-1)$

3 a

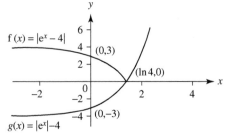

b

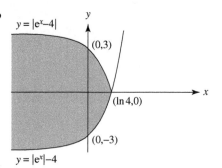

4 a Distance between $(0, 0)$ and (x, y) is

    $D = \sqrt{(y-0)^2 + (x-0)^2}$

      $= \sqrt{x^2 + y^2}$

    $y = 4x + 3 \Rightarrow D = \sqrt{x^2 + (4x+3)^2}$

    $\therefore D = \sqrt{17x^2 + 24x + 9}$

  b i $\left(-\dfrac{12}{17}, \dfrac{3}{17}\right)$ **ii** $\dfrac{3\sqrt{17}}{17}$

5 a i Domain of $f(x)$ $-1 \le x \le 1$

    Range of $f(x)$ $-\dfrac{\pi}{2} \le y \le \dfrac{\pi}{2}$

    Domain of $g(x)$ $-1 \le x \le 1$

    Range of $g(x)$ $0 \le y \le \pi$

    ii

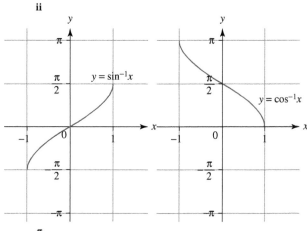

  b $\dfrac{\pi}{2}$

  c Let $\theta = \sin^{-1} x$

    Then $x = \sin\theta$

    So $\cos(2\sin^{-1} x) = \cos 2\theta$

                   $\equiv 1 - 2\sin^2\theta$

                   $= 1 - 2x^2$

6 a 2, (1, 2)

  b i $y = 4$ **ii** $2\sqrt{21}$ **iii** 20.8 (3 sf)

7 68

8

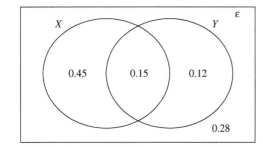

9 No. You would expect roughly $270 \div 6 = 45$
of each outcome, but 72 is very much more than expected.

10 a $X \sim B(37\,890, 0.3)$

b Np is very large.

p is not too far from 0.5

$Y \sim N(37\,890 \times 0.3, 37\,890 \times 0.3 \times 0.7)$

$= N(11\,367, 7956.9)$

c $P(11\,000 \leq X \leq 11\,500)$

$\approx P(10\,999.5 \leq Y \leq 11\,500.5)$

$= P(Y \leq 11\,500.5) - P(Y \leq 10\,999.5)$

$= P\left(Z \leq \dfrac{11\,500.5 - 11\,367}{\sqrt{7956.9}}\right)$

$\quad - P\left(Z \leq \dfrac{10\,999.5 - 11\,367}{\sqrt{7956.9}}\right)$

$= 0.93275 - 0.0000189 \approx 0.9327$

11 a 0.07 **b** 0.26 **c** 0.23

d

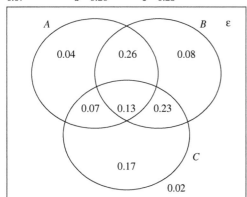

12 a Continuous. The data has been rounded for presentation but weight takes continuous values.

b 1057 g, 1173 g

c 1018 g, 1174 g

d

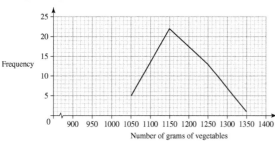

13 a i 1645.5 **ii** 111

b i Upper quartile $+ 1.5 \times$ IQR *or* lower quartile $- 1.5 \times$ IQR

ii No outliers

c Value at $(1104, 545)$ circled.

d i Line through two calculated points (show calculations for point)

ii Strong negative

iii 620 ml (approx.)

iv It is just a prediction; the relationship is not exact.

e i $H_0 : \rho = 0$

$H_1 : \rho \neq 0$

where ρ is the population correlation coefficient.

ii $-0.934 < -0.553$

The result is significant at the 5% level.

There is sufficient evidence to suggest that there is correlation between the amounts of low- and non-low calorie soft drinks purchased per person per week.

Paper 1 (Set B)

1 $x < y, \ x \leq -y$

2 $y = \dfrac{1}{2}(x+3)$

3 $\dfrac{dy}{dx} = 16(3-5x)^2(4x-3)^3 - 10(4x-3)^4(3-5x)$

$= 2(3-5x)(4x-3)^3 \left[8(3-5x) - 5(4x-3)\right]$

$= 2(3-5x)(4x-3)^3(39-60x)$

$= 6(3-5x)(13-20x)(4x-3)^3$

4 a

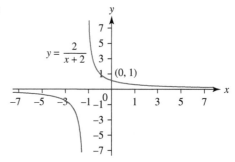

b

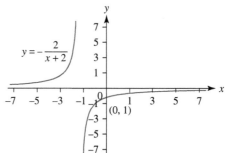

c

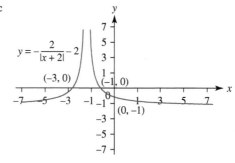

5 a **Method 1**: Substitution

$$x = t\left(\frac{1}{1+t^2}\right) = ty$$

$$\Rightarrow t = \frac{x}{y}$$

Substituting into $y = \frac{1}{t^2+1}$,

$$y = \frac{1}{\left(\dfrac{x}{y}\right)^2 + 1}$$

$$= \frac{y^2}{x^2 + y^2}$$

$$\Rightarrow x^2 + y^2 - y = 0$$

Method 2: Combining x and y directly

$$x^2 + y^2 = \frac{1+t^2}{(1+t^2)^2}$$

$$= \frac{1}{1+t^2}$$

$$= y$$

$$\Rightarrow x^2 + y^2 - y = 0$$

b $1 + (x-3)^2 = \dfrac{y^2}{16}$

6 a $\dfrac{\mathrm{d}T}{\mathrm{d}t}$ is the rate of change of temperature.

k is the constant of proportionality.

k is negative because the temperature is decreasing.

$(T - T_0)$ is the difference between the temperature of the body and the temperature of the surroundings.

b i $\dfrac{1}{10}\ln 2$ **ii** $35°C$ **iii** 25 minutes and 51 seconds

c T will converge to $20°C$.

The tea will cool to the temperature of its surroundings.

7 a $27^2 = 729$, $9^3 = 729$

b $50 = 3 + 21 + 36$ (or equivalent eg $1 + 21 + 28$ etc.)

$51 = 6 + 45$ (or equivalent)

$52 = 1 + 6 + 45$ (or equivalent)

$53 = 10 + 15 + 28$ (or equivalent)

$54 = 3 + 6 + 45$ (or equivalent)

$55 = 55$ (or equivalent)

$56 = 1 + 55$ (or equivalent)

$57 = 1 + 1 + 55$ (or equivalent)

$58 = 3 + 55$ (or equivalent)

$59 = 1 + 3 + 55$ (or equivalent)

$60 = 15 + 45$ (or equivalent)

Fiona's claim is correct.

c Assume the contrary, i.e. the number of primes is finite.

Let the primes be called P_1, P_2, P_3, ... , P_n

Let a new number, Q be equal to the product of **all** of the primes, plus 1

i.e. $Q = (P_1\ P_2\ P_3\ ...\ P_n) + 1$

There are only two possibilities:

1. Q is prime

and since $Q \neq P_1$, P_2, P_3, ... , P_n, this contradicts our original assumption that the only primes were P_1, P_2, P_3, ... , P_n

2. Q is not prime.

By the principle of prime factorisation, Q must be divisible by a prime number P_1, P_2, P_3, ... , P_n

However, dividing Q by any of P_1, P_2, P_3, ... , P_n leaves a remainder of 1

So Q must be divisible by a different prime, not contained in P_1, P_2, P_3, ... , P_n This contradicts our original assumption that the only primes were P_1, P_2, P_3, ... , P_n

Both arguments result in a contradiction.

$\Rightarrow$ The number of primes is infinite.

8 a i $x\ln x - x + c$

ii Let $u = \ln x \Rightarrow \dfrac{\mathrm{d}u}{\mathrm{d}x} = \dfrac{1}{x}$

$$\frac{\mathrm{d}v}{\mathrm{d}x} = \ln x \Rightarrow v = x\ln x - x$$

$$\int (\ln x)^2\,\mathrm{d}x = (x\ln x - x)\ln x - \int (\ln x - 1)\,\mathrm{d}x$$

$$= (x\ln x - x)\ln x - (x\ln x - x - x) + d$$

$$= x(\ln x)^2 - 2x\ln x + 2x + d$$

$$= x(\ln x)^2 + 2x(1 - \ln x) + d$$

b $3 - e$

9 a $2\sin\left(\theta + \dfrac{\pi}{3}\right)$ **b** $\dfrac{-23\pi}{12}, \dfrac{-11\pi}{12}, \dfrac{\pi}{12}, \dfrac{13\pi}{12}$

10 a $2 - \dfrac{x}{192} - \dfrac{5x^2}{147456}$ **b** $|x| < 64$ **c** 1.9948

11 a

$$\ln y = \ln\left[\frac{x^{\frac{2}{3}}(6x+1)^4}{\sqrt{x^2+3}}\right]$$

$$= \ln x^{\frac{2}{3}} + \ln(6x+1)^4 - \ln(x^2+3)^{\frac{1}{2}}$$

$$= \frac{2}{3}\ln x + 4\ln(6x+1) - \frac{1}{2}\ln(x^2+3)$$

$\therefore 6\ln y = 4\ln x + 24\ln(6x+1) - 3\ln(x^2+3)$

b $\dfrac{6}{y}\dfrac{\mathrm{d}y}{\mathrm{d}x} = \dfrac{4}{x} + \dfrac{144}{6x+1} - \dfrac{6x}{x^2+3}$

$$\frac{\mathrm{d}y}{\mathrm{d}x} = y\left(\frac{2}{3x} + \frac{24}{6x+1} - \frac{x}{x^2+3}\right)$$

$$= \frac{x^{\frac{2}{3}}(6x+1)^4}{\sqrt{x^2+3}}\left(\frac{2}{3x} + \frac{24}{6x+1} - \frac{x}{x^2+3}\right)$$

12 a $x^{\frac{1}{12}}$ **b** $\dfrac{2x-1}{x}$ **c** $x^2 - 5x$

Paper 2 (Set B)

1 4

2 $\dfrac{\ln 60}{2\ln 3}$

3 8

4 $p = q - 1$

5 $y - 2x + 3 = 0$

6 a translation with vector $\begin{pmatrix} 4 \\ 3 \end{pmatrix}$

b i Domain $x \geq 4$, Range $\mathrm{f}(x) \geq 3$

ii

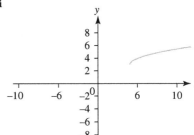

c i $(x-3)^2 + 4$, domain $x \geq 3$, range $\mathrm{f}^{-1}(x) \geq 4$

ii

7 a $-2 < x < 2$

b Since $f''(x)$ changes sign as it passes through $x = 0$, there is a point of inflection at $x = 0$

$f(0) = 3 \Rightarrow$ inflection point at $(0, 3)$

8 $-\sin\theta$

9 $\ln\left(\dfrac{C(x-1)}{x}\right)$

10 $11\mathbf{i} - 2\mathbf{j}$

11 $20.8\,\text{N}$

12 $T = 49.6\,\text{N}, R = 65.3\,\text{N}$

13 a $8.59\,\text{m}$

b $17\,\text{m s}^{-1}$, $30°$ below horizontal (both to 2 sf)

14 a

5.8v N

695.8 N or 71g N

b $120\,\text{m s}^{-1}$

c Whilst the skydiver is still accelerating,

$$\mathbf{F} = m\mathbf{a}$$

$$696 - 5.8v = 71\frac{dv}{dt}$$

$$5.8(120 - v) = 71\frac{dv}{dt}$$

$$120 - v = 12.2\frac{dv}{dt}$$

d $17\,\text{s}$ (to 2 sf)

15 a 1.43 **b** $5.3\,\text{N}$ (to 2 sf), $0.89\,\text{m s}^{-2}$ (to 2 sf)

Paper 3 (Set B)

1 x^4

2 $2x \cos x^2$

3 a $C = \dfrac{5}{9}F - \dfrac{160}{9}$

b p = Temperature change in Celsius for every $1°$ temperature change in Fahrenheit

q = Temperature in Celsius when temperature in Fahrenheit is zero

c $C = 160°$

4 a $x > \dfrac{20}{19}$ **b** $-\dfrac{3}{4} \le x \le \dfrac{5}{2}$ **c** $-2 < x < -1, \ 1 < x < 2$

5 a

x	2	3	5	6	8
y	7.68	6.144	3.932 16	3.145 728	2.013 265 92
$\log_{10} x$	0.301 030	0.477 121	0.698 970	0.778 151	0.903 090
$\log_{10} y$	0.885 361	0.788 451	0.594 631	0.497 721	0.303 901

b i $\log_{10} y$

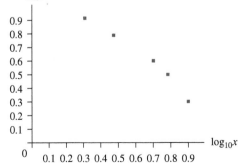

ii $\log_{10} y$

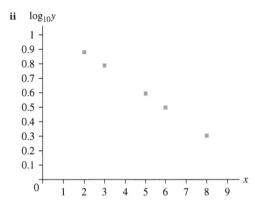

c i Graph of $\log_{10} y$ against x gives a straight line, so data obeys this relationship.

ii 12 (to 2 sf), 0.8 (to 1 sf)

6 Left hand side gives

$$\frac{\sin^2 2x}{4} + \sin^4 x$$

$$= \frac{(2\sin x \cos x)^2}{4} + \sin^4 x$$

$$= \sin^2 x \cos^2 x + \sin^4 x$$

$$= \sin^2 x(\cos^2 x + \sin^2 x)$$

$$= \sin^2 x$$

Right hand side gives

$$1 - \frac{\cot^2 x}{\operatorname{cosec}^2 x}$$

$$= 1 - \frac{1}{\operatorname{cosec}^2 x}(\operatorname{cosec}^2 x - 1)$$

$$= 1 - \sin^2 x\left(\frac{1}{\sin^2 x} - 1\right)$$

$$= 1 - (1 - \sin^2 x)$$

$$= \sin^2 x$$

$$= \text{Left hand side}$$

7 a $(0,0)$

$y = 6(t+2)(t-2) = 0$ for two different values of t i.e. $t = 2$ and $t = -2$

For both values, $x = 0$

$\therefore$ Curve crosses itself at $(0,0)$

b $y = 3x$ and $y = -3x$

8 0.6

9 a 0.4207 **b** 0.6554 **c** 0.2347

10 a i 0.3 **ii** 0.564 **iii** 0.226

b $P(\text{green} \mid \text{plain}) = \dfrac{50}{27 + 23 + 50} = 0.5$

$P(\text{green}) = \dfrac{75 + 50}{250} = 0.5$

These are equal, so the events are independent.

11 a $P(X = 0) = P(X = 4) = \dfrac{k}{9}$

$P(X = 1) = P(X = 3) = \dfrac{k}{3}$

$P(X = 2) = k$

$2 \times \dfrac{k}{9} + 2 \times \dfrac{k}{3} + k = 1$

$k = \dfrac{9}{17}$

b i Y follows a continuous distribution, so a continuity correction must be used.

ii $P(X = 0 \text{ or } 4) = 0.022 32$, $P(X = 1 \text{ or } 3) = 0.2297$, $P(X = 2) = 0.4950$

c It reflects the right shape of the distribution in terms of being symmetric about 2, but the values are too different.

12 a Stratified – When the population can be divided into separate groups which you expect to be represented differently for the survey, you sample from each group in proportion to its size in the population.

Cluster – Split the population into groups (clusters) which you expect to be similar, then sample from some chosen clusters.

b The Postcode Address File and the Land and Property Services Agency list.

c Perhaps only people with the time to spare might agree to take the survey. You might exclude people who are very busy, or elderly people, who could easily have different eating habits.

13 a i, iii

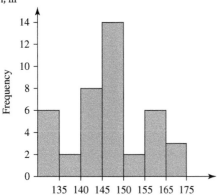

Mass of fish eaten per person per week (g)

ii $115 \le m < 135$ occurs with frequency 6, $135 \le m < 140$ occurs with frequency 2

b The graph is symmetric...

...about a clear mean value.

However, there are fewer outcomes than you might expect in the $150 \le m < 155$ range.

c 146 g (to 3sf), 131 (to 3 sf)

d 248p, 379

e $H_0 : \mu = 150$

$H_1 : \mu \ne 150$

Where μ is the mean number of grams of fish eaten per person per week.

Either

The test statistic is $\dfrac{146 - 150}{\sqrt{\dfrac{131.2}{41}}} = -2.236$.

The critical value is -1.9599

$-2.1997 < -1.9599$ so the result is significant.

or

The test statistic has a p-value of 1.39%

$1.39\% < 2.5\%$ so the result is significant.

Reject H_0 because...

...there is sufficient evidence to suggest that the average of the average amounts of fish eaten per person per week is not 150g.

Paper 1 (Set C)

1 $\sin\theta < 0$ and $\cos\theta < 0$

2 none of these

3 $x + \ln 2$

4 $\dfrac{3+x}{3+3x}$

5 a The integrand does not exist (or is not defined) when $x = 0$

The area would be infinite, as $\dfrac{1}{x^2} \to \infty$ as $x \to 0$

$\dfrac{1}{x^2}$ not continuous on $[-2, 3]$, and the Fundamental Theorem of Calculus only applies to continuous functions.

Sally is correct.

b $\dfrac{1}{x^2}$ is always positive, so the area underneath its graph should be positive.

However, Mark's calculation gives a negative answer, so must be incorrect.

6 a i 37.651 (to 3 d.p.)

ii

Overestimate.

Curve is concave up, which means that the curve lies completely below the top of each strip. Thus the area under the curve is less than the sum of the areas under each of the strips.

b i Now $(x+3)^2 = x^2 + 6x + 9$

Also, $x(x-1) + a(x-1) + b$

$= x^2 - x + ax - a + b$

$= x^2 + (a-1)x + (b-a)$

For $(x+3)^2 = x(x-1) + a(x-1) + b$,

we need: $a - 1 = 6 \Rightarrow a = 7$

$b - a = 9 \Rightarrow b = 16$

ii $20 + 16\ln 3$

iii 0.2%

7 a $-\dfrac{8}{9}$ **b** 1   **c** $\dfrac{1}{2}$

8 a $x^2 + y^2 + 8x - 6y - 11 = 0$

$\Rightarrow (x+4)^2 - 16 + (y-3)^2 - 9 - 11 = 0$

$\Rightarrow (x+4)^2 + (y-3)^2 = 36$

$\Rightarrow$ Centre $(-4, 3)$ and radius 6

$x^2 + y^2 - 16x - 16y + 79 = 0$

$\Rightarrow (x-8)^2 - 64 + (y-8)^2 - 64 + 79 = 0$

$\Rightarrow (x-8)^2 + (y-4)^2 = 49$

$\Rightarrow$ Centre $(8, 8)$ and radius 7

Distance between centres

$= \sqrt{(8-3)^2 + (8--4)^2}$

$= 13$

Sum of the two radii $= 6 + 7 = 13 =$ Distance between centres

$\therefore$ Circles touch.

8 b i Point of contact is $\dfrac{6}{13}$ of the way along the line segment from $(-4, 3)$ to $(8, 8)$

ie $\left[-4 + \dfrac{6}{13}(8-(-4)), \ 3 + \dfrac{6}{13}(8-3)\right] = \left(\dfrac{20}{13}, \dfrac{69}{13}\right)$

ii $y - \dfrac{69}{13} = -\dfrac{12}{5}\left(x - \dfrac{20}{13}\right)$

9 a $\dfrac{dP}{dt} = kP$

b $\displaystyle\int \dfrac{dP}{P} = \int k\,dt$

$\Rightarrow \ln P = kt + \ln A$

$\Rightarrow P = e^{kt}e^{\ln A}$

$P = Ae^{kt}$

c 5.2×10^6, $\dfrac{1}{14}\ln\left(\dfrac{15}{13}\right)$, 8.15×10^6 (to 3 s.f.)

10 $4(x+2)^2 + 25(y+1)^2 = 100$

11 a Letting $u = \cos x$ gives

$$\frac{du}{dx} = -\sin x \Rightarrow dx = -\frac{du}{\sin x}$$

Hence

$$\int \tan x \, dx = \int \frac{\sin x}{\cos x} dx$$

$$= \int -\frac{\sin x}{u} \frac{du}{\sin x}$$

$$= \int -\frac{1}{u} du$$

$$= -\ln|u| + c$$

$$= -\ln|\cos x| + c$$

$$= \ln\left|(\cos x)^{-1}\right| + c$$

$$= \ln|\sec x| + c$$

b Method 1: Letting $u = \sec x$

Then $\dfrac{du}{dx} = \sec x \tan x \Rightarrow dx = \dfrac{du}{\sec x \tan x}$

$$\int \sec^4 x \tan x \, dx = \int \sec^3 x (\sec x \tan x) \, dx$$

$$= \int u^3 (\sec x \tan x) \frac{du}{(\sec x \tan x)}$$

$$= \int u^3 \, du$$

$$= \frac{u^4}{4} + c$$

$$= \frac{\sec^4 x}{4} + c$$

OR Method 2: Letting $u = \tan x$

$$\frac{du}{dx} = \sec^2 x \Rightarrow dx = \frac{du}{\sec^2 x}$$

$$\int \sec^4 x \tan x \, dx$$

$$= \int \sec^2 x \tan x (\sec^2 x) \, dx$$

$$= \int (1 + \tan^2 x) \tan x (\sec^2 x) \, dx$$

$$= \int (1 + u^2) u (\sec^2 x) \frac{du}{(\sec^2 x)}$$

$$= \int (1 + u^2) u \, du$$

$$= \int u + u^3 \, du$$

$$= \frac{u^2}{2} + \frac{u^4}{4} + k$$

$$= \frac{\tan^2 x}{2} + \frac{\tan^4 x}{4} + k$$

$$= \frac{\sec^2 x - 1}{2} + \frac{(\sec^2 x - 1)^2}{4} + k$$

$$= \frac{2\sec^2 x - 2}{4} + \frac{\sec^4 x - 2\sec^2 x + 1}{4} + k$$

$$= \frac{\sec^4 x - 1}{4} + k$$

$$= \frac{\sec^4 x}{4} + C, \text{ where } C = k - \frac{1}{4}$$

c $\displaystyle\int \left[\sec x \tan x (\sec^3 x) - \sec x + \tan x \right] dx$

$$= \int \sec^4 x \tan x \, dx - \int \sec x \, dx + \int \tan x \, dx$$

[Method 1 above]

$$\frac{1}{4}\sec^4 x - \ln(\sec x + \tan x) + \ln(\sec x) + k$$

$$= \frac{\sec^4 x - 1}{4} - \ln(\sec x + \tan x) + \ln(\sec x) + \ln A$$

[Method 2 above]

$$= \frac{\sec^4 x - 1}{4} - \ln|\sec x + \tan x| + \ln|\sec x| + \ln A$$

(where $\ln A$ is an arbitrary constant)

$$= \frac{\sec^4 x - 1}{4} + \ln\left(\frac{A\sec x}{\sec x + \tan x}\right)$$

12 a $20x^4 - 72x^3 + 87x^2 - 38x + 3$

$$= (10x^2 - 11x + 1)(ax^2 + bx + c)$$

Equating coefficients of x^4:

$$20 = 10a \Rightarrow a = 2$$

Equating constants:

$$3 = c$$

Equating coefficients of x^3:

$$-72 = 10b - 11a \Rightarrow b = -5$$

$$\therefore 20x^4 - 72x^3 + 87x^2 - 38x + 3$$

$$= (10x^2 - 11x + 1)(2x^2 - 5x + 3)$$

$$\Rightarrow (10x^2 - 11x + 1) \text{ is a quadratic factor}$$

b $(10x - 1)(2x - 3)(x - 1)^2$

c

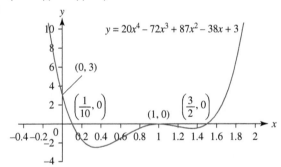

d Reflection in the y-axis.

13 a i $\dfrac{3}{n^2}\left[\dfrac{n(n+1)}{2}\right] + 4 = \dfrac{3(n+1)}{2n} + 4$

$$= \frac{3}{2} + \frac{3}{2n} + 4$$

$$= \frac{11}{2} + \frac{3}{2n}$$

ii $S_n = \dfrac{n}{2}\left[2(1) + (n-1)(1)\right]$

$$S_n = \frac{n}{2}[2 + n - 1]$$

$$S_n = \frac{n(n+1)}{2}$$

b i $c_i = 0 + \left(\dfrac{1-0}{n}\right)i \Rightarrow c_i = \dfrac{i}{n}$

ii $\dfrac{11}{2}$

iii $\dfrac{11}{2}$

Paper 2 (Set C)

1 $|x| < \dfrac{b}{a}$

2 a $\dfrac{1}{4}x^{-2}e^{\frac{1}{2}x}(x-2)$

 b $\ln|\ln x| + c$

3 $12 - 4\sqrt{3}$ units, $10 - 4\sqrt{3}$ units

4 $2x^2 - 12x + 19 = 2(x^2 - 6x) + 19$

$$= 2\left[(x-3)^2 - 9\right] + 19$$

$$= 2(x-3)^2 + 1 \ (x-3)^2 \geq 0 \Rightarrow 2(x-3)^2 + 1 > 0$$

5 a Let $f(x) = x^3 - x^2 + 3x - 4$

$f(1) = -1 < 0$

$f(1.5) = 1.625 > 0$

$f(x)$ is **continuous** and there is a **change of sign**.

$\Rightarrow 1 < \alpha < 1.5$

 b $x^3 = x^2 + 4 - 3x$

$x^2 = \dfrac{x^2 + 4}{x} - 3$

$x = \sqrt{\dfrac{x^2 + 4}{x} - 3}$

 c $x_{n+1} = \sqrt{\dfrac{x_n^2 + 4}{x_n} - 3}$, 1.222(to 3 dp)

 d

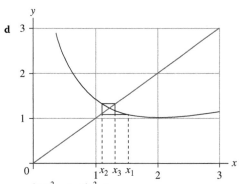

6 $\dfrac{2(xy^2 - x - \sin^2 y)}{4(x + \sin^2 y)\sin y \cos y - 2x^2 y}$

7 a i $\dfrac{2\pi}{b}$ **ii** 1

 b i $\dfrac{2\pi}{b}$ **ii** a

 c i $\dfrac{2\pi}{b}$ **ii** a

 d $y = 4.75\cos\dfrac{\pi}{6}x + 12.25$

8 Area between graph and x–axis

9 $\dfrac{3}{5}\mathbf{i} - \dfrac{4}{5}\mathbf{j}$

10 a $140\,\text{N}$ **b** $1.6\,\text{m}$

11 $-\dfrac{3}{4}\mathbf{i} + \mathbf{j}\,\text{ms}^{-2}$

12 a $v = \dfrac{ds}{dt} = \dfrac{d}{dt}(te^{5-t})$

$$= e^{5-t} - te^{5-t}$$

$$= -e^{5-t}(t-1)\,\text{ms}^{-1}$$

 b $6e^4\,\text{N}$

13 $2.75\,\text{m}$

14 a $x = ut\cos\theta$

$y = ut\sin\theta - \dfrac{1}{2}gt^2$

$t = \dfrac{x}{u\cos\theta}$

$y = u\dfrac{x}{u\cos\theta}\sin\theta - \dfrac{1}{2}g\left(\dfrac{x}{u\cos\theta}\right)^2$

$= x\tan\theta - \dfrac{1}{2}g\dfrac{x^2}{u^2\cos^2\theta}$

$y = x\tan\theta - \dfrac{g\sec^2\theta}{2u^2}x^2$

 b $74°$ (to 2 sf), $30°$ (to 2 sf)

 c The minimum angle would be greater.
 The maximum angle would be smaller.
 Air resistance would mean that the horizontal range of the jump would be less for each value of θ

Paper 3 (Set C)

1 $2^x \ln 2$

2 $x = e^{kt+c}$ or $x = Ae^{kt}$

3 $-\csc 2y$

4 Yes, as his solution is equivalent to $-a < x < a$

5 a

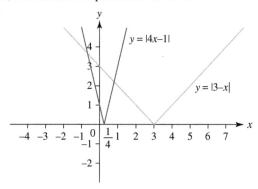

 b $\left(-\dfrac{2}{3}, \dfrac{11}{3}\right)$, $\left(\dfrac{4}{5}, \dfrac{11}{5}\right)$

 c $-\dfrac{2}{3} < x < \dfrac{4}{5}$

6 $\dfrac{3}{8}$

7 Let the two integers be A and B
 Then the two-digit numbers are
 AB and BA
 Adding, we have
 $10A + B + 10B + A$
 $= 11A + 11B$
 $= 11(A + B)$
 which is a multiple of 11
 $\therefore$ Vince's observation is always true.

8 Translation with vector $\begin{pmatrix} -\dfrac{\pi}{12} \\ 0 \end{pmatrix}$

9 a $2, \sqrt{2}, 1$ **b** $\dfrac{1}{8} + \dfrac{1}{8}\sqrt{2}$

10 a $\sin\left(\dfrac{\pi}{2} - A - B\right) = \sin\left(\dfrac{\pi}{2} - A\right)\cos B - \cos\left(\dfrac{\pi}{2} - A\right)\sin B$

$$= \cos A\cos B - \sin A\sin B$$

 But $\sin\left(\dfrac{\pi}{2} - A - B\right)$

$$= \sin\left[\dfrac{\pi}{2} - (A + B)\right] = \cos(A + B)$$

$\therefore \cos(A + B) = \cos A\cos B - \sin A\sin B$

10 b Replacing B by A in **a**, we have

$\cos(A+A) = \cos A \cos A - \sin A \sin A$

$\cos 2A = \cos^2 A - \sin^2 A$

Using $\sin^2 A = 1 - \cos^2 A$, we get

$\cos 2A = \cos^2 A - (1 - \cos^2 A)$

$\Rightarrow \cos 2A = 2\cos^2 A - 1$

$\Rightarrow \cos^2 A = \dfrac{1 + \cos 2A}{2}$

c $\dfrac{1}{32}(12A + 8\sin 2A + \sin 4A) + c$

11 $P(B \mid A) = P(A)$

12 $H_1 : \mu > 7$

13 a (very) weak positive

b $H_0 : \rho = 0$

$H_1 : \rho \neq 0$

where ρ is the population correlation coefficient between the scores on the two tests.

Since $0.2684 < 0.514\ldots$

$\ldots$there is insufficient evidence to reject H_0.

You conclude that there is no correlation between scores on the two tests.

14 a The events are mutually exclusive and their probabilities sum to 1

b

EEEE	EEEO	EEOE	EEOO
EOEE	EOEO	EOOE	EOOO
OEEE	OEEO	OEOE	OEOO
OOEE	OOEO	OOOE	OOOO

c $\dfrac{5}{16}$

d $\dfrac{5}{11}$

15 a Only those who are available in the middle of the day will take part, e.g. the unemployed. This could bias the sample.

b $X \sim \text{B}\left(30, \dfrac{1}{3}\right)$

c $10, \dfrac{20}{3}$

d 0.0355 (to 3 sf)

e $H_0 : p = \dfrac{1}{3}$

$H_1 : p > \dfrac{1}{3}$

where p is the probability that a randomly chosen person supports the policy.

Either

Since $P(X \geq 14) = 8.98\%$

and $P(X \geq 15) = 4.35\%$,

The critical region is $X \geq 15$

The result (15 people) is contained within the critical region.

Or

Half the people is 15 supporting the policy.

$P(X \geq 15) = 4.34\% < 5\%$

Reject H_0

There is sufficient evidence at the 5% level to suggest that the company's assumption was an under-estimate.

16 a 42

b 24 (to 2 sf)

c $\dfrac{27}{41}$

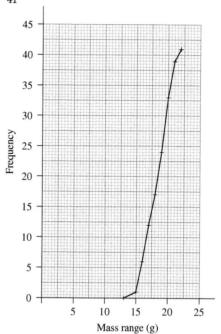

d

e The central tendency of the coffee is much lower but so is the spread.

17 a

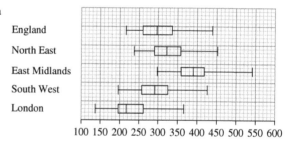

b Largest: East Midlands, Smallest: North East

c IQR $= 59$

$UQ + 1.5 \times 59 = 506.5$

$LQ - 1.5 \times 59 = 270.5$

There is at least one outlier as the maximum (542) is an outlier.

Don't exclude as there's no reason to believe it isn't a real value.

d The South West has the most similar 5–number summary values to England as a whole.

e Advantage: It is easier and cheaper to take a sample.

Disadvantage: Could succumb to bias from regional variations.

f A simple random sample is one in which every possible sample of the desired size is equally likely to be chosen.

Cluster sampling definitely excludes all entries not contained within a cluster, and so not every member of the population has a chance of being chosen.